Needs and Feasibility

A Guide for Engineers in Community Projects

The Case of Waste for Life

Needs and Feasibility: A Guide for Engineers in Community Projects — The Case of Waste for Life
Caroline Baillie, Eric Feinblatt, Thimothy Thamae, and Emily Berrington

ISBN: 978-3-031-79957-0 paperback
ISBN: 978-3-031-79958-7 ebook

DOI 10.1007/978-3-031-79958-7

A Publication in the Springer series
SYNTHESIS LECTURES ON ENGINEERS, TECHNOLOGY, AND SOCIETY

Lecture #13
Series Editor: Caroline Baillie, *University of Western Australia*
Series ISSN
Synthesis Lectures on Engineers, Technology, and Society
Print 1947-3633   Electronic 1947-3461

Synthesis Lectures on Engineers, Technology, and Society

Editor
Caroline Baillie, *University of Western Australia*

The mission of this lecture series is to foster an understanding for engineers and scientists on the inclusive nature of their profession. The creation and proliferation of technologies needs to be inclusive as it has effects on all of humankind, regardless of national boundaries, socio-economic status, gender, race and ethnicity, or creed. The lectures will combine expertise in sociology, political economics, philosophy of science, history, engineering, engineering education, participatory research, development studies, sustainability, psychotherapy, policy studies, and epistemology. The lectures will be relevant to all engineers practicing in all parts of the world. Although written for practicing engineers and human resource trainers, it is expected that engineering, science and social science faculty in universities will find these publications an invaluable resource for students in the classroom and for further research. The goal of the series is to provide a platform for the publication of important and sometimes controversial lectures which will encourage discussion, reflection and further understanding.

The series editor will invite authors and encourage experts to recommend authors to write on a wide array of topics, focusing on the cause and effect relationships between engineers and technology, technologies and society and of society on technology and engineers. Topics will include, but are not limited to the following general areas; History of Engineering, Politics and the Engineer, Economics , Social Issues and Ethics, Women in Engineering, Creativity and Innovation, Knowledge Networks, Styles of Organization, Environmental Issues, Appropriate Technology.

Needs and Feasibility: A Guide for Engineers in Community Projects — The Case of Waste for Life
Caroline Baillie, Eric Feinblatt, Thimothy Thamae, and Emily Berrington
2010

Humanitarian Engineering
Carl Mitcham and David Munoz
2010

Engineering and Sustainable Community Development
Juan Lucena, Jen Schneider, and Jon Leydens
2010

Needs and Feasibility
A Guide for Engineers in Community Projects
The Case of Waste for Life

Caroline Baillie
University of Western Australia

Eric Feinblatt
State University New York

Thimothy Thamae
Queens University, Canada

Emily Berrington
Kings College, London

SYNTHESIS LECTURES ON ENGINEERS, TECHNOLOGY, AND SOCIETY #13

ABSTRACT

Needs and Feasibility: A Guide for Engineers in Community Projects—The Case of Waste for Life is the story of Waste for Life (WFL). WFL is a not-for-profit organization that works to promote poverty-reducing solutions to environmental problems, and its educational branch is an international consortium of universities in six countries, involving students in support of community development projects. WFL currently works in Lesotho and Argentina. We present the story of the development of WFL in each country as a case-based guide to engineers, professors and students interested in community development work, particularly in contexts very different from their own. We focus mainly on the set-up stages, framing the projects to ensure that community needs are adequately articulated and acted upon. We begin with needs assessment, what is it that needs to be done – for whom and why? How feasible is this, technically, economically, and can we guarantee sustainability? Before we can decide any of this, we need to understand and map the territory – who are the key players, who have the most influence, and who will be most impacted by what we are doing? What is the role of the local government? If the groups are working as cooperatives, what does this mean, and what are these groups looking for? What is the technical solution going to look like? If it is a product, how will it be marketed? What other social, environmental, and economic impacts will it have and on whom? Once these have all been negotiated, and it is clear that all parties are working towards a mutually acceptable goal, how do we move forward so that any dependence on external partners is removed? When do we bring students into the work? What role can they play? Should they stay at home and support the project from there or is it better to do work in the field? This guide will be useful for the student engineer or the experienced engineer or professor who is interested in moving towards socially just engineering development work but has no idea where to begin. The real difficulties and on the ground issues encountered by the Waste for Life team are presented honestly and with the knowledge that we must learn from our mistakes. Only then can we hope to gain a better understanding of our potential role in supporting community development and move towards a better future.

KEYWORDS

waste, cartoneros, urban recoverer, recycling, Buenos Aires, Lesotho, community project, development, capacity building, plastic, natural fibre, composites, Argentina, engineering

Contents

Acknowledgments

We would like to dedicate the book to the cartoneros in Buenos Aires and to the cooperatives in Lesotho, without whom none of our work would be possible or purposeful. We would also like to thank all of our friends in Buenos Aires and Lesotho for their time working with us and for their belief in Waste for Life. We are indebted to Darko Matovic for his hotpress design and continued work with WFL. A thank you to the students and their professors at Rhode Island School of Design for their marketing analysis, Nils Rehmann for his work on the cost benefit template, to the Way Memorial Trust Fund at Queens University for supporting our visits to Lesotho, and a big thank you to everyone we interviewed and to all the students, engineers, artists, architects, and professors around the world who have donated their time to support WFL.

Caroline Baillie, Eric Feinblatt, Thimothy Thamae, and Emily Berrington
February 2010

Preface

Engineers can help make the world a better place. We can provide better access to water, health, food, shelter, education and warmth, and we have the potential to help make structural changes that will change people's lives. For better or for worse. This book series is intended to support those engineers who are interested in the former, and whose praxis contributes to redefining what 'better' means. We explore the notion of social justice and how it relates to engineering, how engineers can work to promote sustainable, mutually negotiated programs of work. This book complements its partner in the series, *Engineering and Sustainable Community Development* by Lucena et al, 2010, (which gives an overview of the critical issues involved in engineering development work), by presenting the story of Waste for Life. WFL is a not-for-profit organisation that works to promote poverty-reducing solutions to environmental problems, and its educational branch is an international consortium of universities in six countries, involving students in support of community development projects. WFL currently works in Lesotho and Argentina. We present the story of the development of WFL in each country as a case-based guide to engineers, professors and students interested in community development work, particularly, in contexts very different from their own. We focus mainly on the set-up stages, framing the projects to ensure that community needs are adequately articulated and acted upon. It is rarely the case that a perfect project is handed to us, with complete trust having been established with local partners so we can feel confident that our contribution achieves the best possible balance between facilitation and intervention. Even NGOs and government bodies can misrepresent the needs of the local communities we might work with, so the ground work that needs to be done before any project can take place is very important. We consider the needs assessment, what is it that needs to be done – for whom and why? How feasible is this, technically, economically, and to ensure sustainability? Before we can decide any of this, we need to understand and map the territory – who are the key players, who have the most influence and who will be most impacted by what we are doing? What is the role of the local government? If the groups are working as cooperatives, what does this mean and what are these groups looking for? What is the technical solution going to look like? If it is a product, how will it be marketed? What other social, environmental and economic impacts will it have and on whom? Once these have all been negotiated and it is clear that all parties are working towards a mutually acceptable goal, how do we move forward so that any dependence on external partners is removed? When do we bring students into the work? What role can they play? Should they stay at home and support the project from there or is it better to do work in the field? This guide will be useful for the student engineer or the experienced engineer or professor who is interested in moving towards socially just engineering development work but has no idea where to begin. The real difficulties and on the ground issues encountered by the Waste for Life team are presented honestly and with the knowledge that we must learn from our mistakes. Only then can we hope to gain a better understanding of our potential role in supporting community development and move towards a better future.

Caroline Baillie, Eric Feinblatt, Thimothy Thamae, and Emily Berrington
January 2010

CHAPTER 1

An Introduction

Hovering above any major city in an aircraft, one can look down and view all the arterial roads, railways, and canals leading towards and away from that city—its lifeblood. The city is the heart, and the trucks, trains and boats are the blood, pumping the goods to where they are needed. They are the in and out valves of society. Then the thought arises, this could all stop. In fact, if engineers laid down their tools and their calculators today, everything would soon stop. Nothing would work, nothing would move. This thought can make you feel very powerful or very responsible, depending on your personality. It is assumed that if you are reading his book, you feel the latter.

Certainly, engineers have the potential to help make the world a better place, but they can also do the opposite. And so far, alongside improving the standard of living of a great many people, we have also done a very good job at harming our planet and changing irreparably the way many people live, for the worse. Many of us live at a faster pace than we would prefer and sit behind computers many hours a day. There are some who say that this was not inevitable but that we have allowed technological progress to rule the way we live. These are the technological determinists. Others suggest that it is the inverse, that society dictates what technology develops, that market forces have created the need for more and more gadgets to help us do things faster and more efficiently, often neglecting the consequences. What is certain, however, is that as engineers, we have the technology and knowledge to move societies and communities in different directions, though what usually happens is that we follow the demands of the politicians, the CEOs, and the economists. If we are true professionals, then key decisions relating to technological development ought to involve engineers. We should contribute to the policies that will ultimately determine what we do, how we do it, and how what we do is used. But in order to do that we need to know what we are getting ourselves into. And what we are getting others into. In this book, we explore the notion of social justice and how it relates to engineering; we also explore how engineers can work to promote sustainable, mutually negotiated 'development'.

In *Engineers in a Local and Global Society* (Baillie, C., 2006) in this series, we discuss the multiple definitions of development that are pertinent to our discussion here and which we summarise. Development is not a simple issue. Because it is a word that is simultaneously construed to serve different and often diametrically opposed purposes, there are many ways of looking at it (Leftwich, A., 2000). Most commonly, development is understood as historical progress tethered to material advancements serving national strategies that promote economic growth. But development can also be tied to social objectives that enhance participatory democratic processes and empower people to gain a measure of autonomous control over their own lives. Market interests, social interests, and national interests often collide under the development umbrella. Development can be modernization, westernization or planned exploitation, management or use of natural resources; it can be increased GNP, the promotion of economic, political and social advancements or far reaching structural change; it can be top down or bottom up, or it can be a little of both. Development can also been seen as a condition – a state that countries find themselves in. There is the idea that some countries are developed and others are moving towards that end point and, in this case, developed countries often think they are in a better position to know what locals want in the country being 'developed'. This is called trusteeship. But whatever our idea of development – one or all of the above, it is clear that there is not one right way of being that we are all moving towards. We might assume that everyone in a 'developing' country might like to live like those of us who live in the US or in Canada or the UK. But this may not ultimately increase the level of freedom, happiness or even length of life. Sen, A. (1999), a Nobel Prize winner in economics has shown that the longevity of African Americans is lower than people living in very poor developing countries – the stress of living at the bottom of the rung has its toll. The definition or definitions one chooses is going to impact the type of development work we practice, but those practices should always be preceded by asking ourselves who we are developing for and who decides what is being developed.

Related to the rapid development of many countries in the world, we have also experienced in the last two decades, increasing global trade and movement of goods across borders. This has been called 'globalisation.' Again, there are many varied views about globalisation and its impact on under-privileged populations. For many, this is the inevitable movement towards an increasingly interconnected world, and it is not a bad thing. Others, particularly those involved in the 'anti-globalisation' movement, are less convinced of the equitable benefits of these recent trends. They are not against internationalisation, which it is often confused with, but they are against the push for developing countries to become market-driven before they are ready for it. They are against the exploitation of labour in developing countries for the benefit of corporations in developed countries. Capra (2002) sums up his views as follows:

> During the last decade of the twentieth century, a recognition grew among entrepreneurs, politicians, social scientists, community leaders, grass roots activists, artists, cultural historians and ordinary women and men from all walks of life that a new world was emerging – a world shaped by new technologies, new social structures, a new economy and a new culture: 'Globalisation.' With the creation of the World Trade Organisation (WTO) in the mid-1990s, economic globalisation, characterized by 'free trade' was hailed by corporate leaders and politicians as a new order that would benefit all nations, providing worldwide expansion whose wealth would trickle down to all. However, it soon became apparent to increasing numbers of environmentalists and grassroots activists that the new economic rules established by the WTO were manifestly unsustainable and were producing a multitude of interconnected fatal consequences and extensive deterioration of the environment, the spread of new diseases and increasing poverty and alienation….in addition to its economic instability, the current form of global capitalism

is ecologically and socially unsustainable and hence not viable in the long run (Capra, 2002, p. 130–157).

Again, there is no single definition of what globalisation is. It can be seen in purely economic terms – changing the volume and flow of trade, capital, and labour across borders. It can be seen as the changing role of the nation state or of spatial relations as a result of technology and communication. It can be seen as the blending or domination of cultures as "the culture being spread around the globe is generated by corporate products and corporate decisions," and he quotes Benjamin Barber "a culture of advertising, software, Hollywood movies, MTV, theme parks and shopping malls hooped together by the virtual nexus of the information superhighway closes down free spaces, such a culture is unquestionably in the process of forging a global something; but whatever it is, that something is not democratic" (Brawley, 0000, p. 29).

So, to extend our arterial metaphor, we find ourselves sitting in the middle of this global superhighway, watching lives rush by, not knowing who is going where or why, or what they are taking with them. There are no traffic laws or regulations that we can discern. Once a global corporation is outside the nation's trade laws, things become very complicated. We look over to our left and we see some people being run over by the trucks that pass. That we designed. No one says anything. No one knows who to say it to.

From this dismal vision, we withdraw and ask ourselves how we can avoid this situation where somehow, whether we know it or not and against our every intention, we caused the harm, increasing poverty and even death of innocent people in a country that we had not even heard of until our project began. How do we even begin to question these things? What position do we take?

In a recent article, we have discussed the engineering traditions, which steadfastly secure our feet in concrete, ensuring that we stand with security and pride but rendering movement rather difficult (Kabo and Baillie, 2009a). Engineering as an enlightenment project is built on such classical definitions as that of Thomas Tredgold on behalf of the Institution of Civil Engineers in 1828, stating that "[engineering is] the art of directing the great sources of power in nature for the use and convenience of man [sic]" (Johnston et al., 2000, p. 26). But which man (or woman)? 180 years later, we see similar definitions, although in some circles the idea of social impact has begun to bubble to the surface. According to Johnston et al. (2000, p. 26) engineering is the following: "A total societal enterprise, with significant influences on all aspects of human life and a major role to play in moving the world towards particular goals." However, Vesilind (2006, p. 283) suggests that engineers have never been very good at this, since "[t]he engineer is sophisticated in creating technology, but unsophisticated in understanding its application. As a result, engineers have historically been employed as hired guns, doing the bidding of both political rulers and wealthy corporations." Overall, the current, or 'common sense' understanding of engineering among, students, practitioners, and even professors is that engineering is mostly associated with efficiency and profit making, and there is usually an inherent belief that technical development always equates to progress. Furthermore, when it comes to working in cultural contexts different from their own, Northern engineers tend to unquestioningly embrace the idea of 'development as good' in contrast to critics such as Ferguson, who believe that development has become a machine that implements technical solutions to problems that are far from technical in nature (Ferguson, J., 1990). Ferguson critiques the idea of 'developers' socially constructing a third world in such a way that the technical solutions the North has to offer seem to be the way to 'progress.' He does not completely reject the idea of people from the North working together with those from the South but he gives certain criteria for this – groups should understand the politics surrounding the particular issues, and these groups should represent a movement toward local empowerment. We would add, 'do no harm to the community,' which is a much more complex statement than at first glance. Western ideas of empowerment and emancipation may not have the same meaning or, perhaps, are in conflict with traditions that have other compelling meanings in foreign contexts. A benign (for us) enabling technology may have consequences that fiddle around with entire belief systems and disrupt patterns of behavior that are reinforced by a society and most of its artefacts – it's culture, politics, education, and on and on. There is no absolute here.

Gramsci identifies the complexities of working with people across cultures by using the word 'hegemony,' which he defines as the common sense or dominant way of seeing/understanding the world within a given community of practice: "[It] is not a single unique conception, identical in time and space. [...] Its most fundamental characteristic is that it is a conception which, even in the brain of one individual, is fragmentary, incoherent and inconsequential, in conformity with the social and cultural position of those masses whose philosophy it is" (Gramsci, A., 1971, p. 149).

Common sense "is used by Gramsci to mean the uncritical and largely unconscious way of perceiving and understanding 'common' in any given epoch" Hoare and Smith (1971, p. 322), and hegemony is "a process of social control that is carried out through the moral and intellectual leadership of a dominant sociocultural class over subordinate groups" (Darder, Baltodano and Torres, 2009, p. 12). Engineers from the North who work to 'develop' communities from the South are particularly susceptible to hegemonic practices. Developing the sensibility of students to critique and construct alternative, non-hegemonic practices is a key challenge. Baillie and Rose (2004, p. 20) suggest that: "it is important to realise that for something to be known, it must fit within the relevant community's paradigm or 'thought collective.'" Fleck coined the phrase 'thought collective' as: "a community of persons mutually exchanging ideas or maintaining intellectual interaction, we will find by implication that it also provides the special 'carrier' for the historical development of any field of thought, as well as for the given stock of knowledge and level of culture. This we have designated thought style" (Fleck, L., 1979, p. 39).

Engineers, especially student engineers, trying to locate their work within different community's own thought collectives, especially when these are not one-dimensional or static, experience a very tricky negotiation. With students of engineering who have no preparation in the social and political sciences this can lead to a continuation of colonial ignorance and exploitation, the very reverse of what was intended.

Fleck warns us that thinking a certain way will "constrain(s) the individual by determining what can be thought in no other way" (Fleck, L., 1979, p. 99). And when two different thought styles collide: "The alien way of thought seems like mysticism. The questions it rejects will often be regarded as the most important ones, its problems as often unimportant or meaningless trivialities" (Fleck, L., 1979, p. 109): The problem is that when one is stuck inside a thought collective:

> The understanding inherent (in a shared practice) is not necessarily one that gives members broad access to the histories or relations with other practices that shape their own practice. Through engagement, competence can become so transparent, logically ingrained, and socially efficacious that it becomes insular: nothing else, no other viewpoint, can even register, let alone create a disturbance or a discontinuity that would spur the history of practice onward. In this way, a community of practice can become an obstacle to learning by entrapping us in its very power to sustain our identity (Wenger, 1998, p. 175).

Understanding this dilemma has led to the need to develop a new approach to engineering service learning that embraces the differences in cultures and communities, understands and looks for tensions within these, and responds dynamically to the consequences of the entanglements that arise. If we are to develop service learning programs within schools of engineering, which embrace complex cultural dynamics, our challenge appears to be to help students learn how to understand and analyse the important questions within the communities in which they will work. They need to move beyond all their various common sense ways of thinking: their Northern, Western, middle class, often white male, as well as their engineering community of practice.

There are a growing number of critics of the engineering common sense view or thought collective, who aim to prepare students for working in the world in a socially just manner (Catalano, G., 2006, 2007; Riley, D., 2008; Baillie, C., 2006), and a developing network of engineers, academics and students who meet to discuss options for education related to 'engineering, social justice and peace' (http://esjp.wikispaces.com/), but very few, if any university, fully embraces this critique.

Ursula Franklin, a very important thinker and Emeritus Professor of Materials Engineering at the University of Toronto, asks, when deciding upon a particular project, not to simply consider benefits and costs but to question 'whose benefits and whose costs?' Her thinking could serve us well when considering involvement in a development project.

Franklin makes a series of recommendations to do this – to pose these questions before beginning a project:

- promote justice

- restore reciprocity (Franklin worries, for example, that communications technology has become non-communications technology – we are decreasing ways of feeding back)

- confer divisible or indivisible benefits (working on ways to improve the environment is an indivisible benefit – everyone benefits)

- favour people over machines

- minimize or maximize disaster

- promote conservation over waste

- favour reversible over irreversible.

She suggests that for any project, we should do our bookkeeping. However, she believes that we need three sets of books: one for the economy, one for people and social impacts, one for environmental accounting, and that we must ask for each of these areas – who benefits and who pays?

Another important thinker in the area is George Catalano a professor of mechanical engineering at Binghamton University. Catalano has reviewed the major US Engineering Codes of Ethics in respect to their consideration of poverty, security, and nature. His table of findings are given below (Table 1.1). He tells us the following:

> Engineering as a value-laden profession seeks to codify ethical behavior with various codes of conduct as put forth by different engineering societies. There are differences among the different codes but there are some striking similarities. The similarities exist in what has not been included in the ethical codes. While each does speak to the importance of holding paramount the public safety, issues associated with the intimate connection between engineering and war industries and terrorism are not discussed. In addition, no code speaks to the challenge of world poverty and the plight of the under-developed world. With one exception, that of ASCE, the challenge of environmental sustainability is completely ignored (Catalano, G., 2006, p. 14).

Catalano has also created a design cycle, which looks very similar to previous versions that you will have already come across (Catalano, G., 2006, p. 31). However, this cycle suggests that not only do we make sure we are not making the situation worse, but that we are in fact improving it. His steps include the following:

Via Positiva. The problem is identified, fully accepted, and broken down into its various components using the vast array of creative and critical thinking techniques which engineers possess. What is to be solved? For whom is it to be solved?

Via Negativa. Reflection on the possible implications and consequences for any proposed solution are explored. What are the ethical considerations involved? The societal implications? The global consequences? The effects on the natural environment?

Via Creativa. The third step refers to the act of creation. The solution is chosen from a host of possibilities, implemented and then evaluated as to its effectiveness in meeting the desired goals and fulfilling the specified criteria.

Via Transfomativa. The fourth and final step asks the following questions of the engineer: Has the suffering in the world been reduced? Have the social injustices that pervade our global village been even slightly ameliorated? Has the notion of a community of interests been expanded? Is the world a kinder, gentler place borrowing from the Greek poet Aeschylus?

If we join together Catalano's design cycle with Franklin's bookkeeping, we have a simple system to use as checks and balances when working on any engineering project, especially development

Table 1.1: Comparison of Attitudes towards Security, Poverty and Nature among Various Engineering Societies (Catalano, G., 2006).

Code of Conduct	Relevant Canons and Principles	Attitudes Towards Security	Attitudes Towards Poverty	Attitudes Towards Nature
NSPE	Hold paramount the safety, health, and welfare of the public	No explicit reference	No explicit reference	No explicit reference
ASME	Uphold and advance the integrity, honor, and dignity by using their knowledge and skill for the enhancement of human welfare	No explicit reference	No explicit reference	No explicit reference
ASCE	Hold paramount the safety, health and welfare of the public and shall strive to comply with the principles of sustainable development	No explicit reference	No explicit reference	Sustainable development linked solely to meeting human needs
IIE	Accept responsibility in decisions consistent with the safety, health and welfare of the public, and to disclose promptly factors that might endanger the public or the environment	No explicit reference	No explicit reference	Endangering environment not explored
IIE (ABET)	Shall hold paramount the safety, health and welfare of the public in the performance of their professional duties	No explicit reference	No explicit reference	No explicit reference
AIChE	Hold paramount the safety, health and welfare of the public and protect the environment	No explicit reference	No explicit reference	Protecting the environment not explored

projects. We have a system which allows us to check to see if what we are planning is sustainable – within the resources (both economic and environmental) available, but also we are checking to see who is benefiting and/or losing out in the new system, and whether this is what we intended. These are critical steps to take in any problem definition stage of a project.

This book is intended to be a partner volume to *Engineering and Sustainable Community Development* in this series by Lucena and Schneider (2010), which presents the case for critiquing engineering development work from a community perspective. In this volume, we demonstrate how these lessons have been applied in a real context and show how messy and ill defined they really are. Decisions are not black and white. Considerations rely on morals and ethics, values and political affiliations (for help with these the authors recommend a visit to Part 2 of *Engineering and Society: Working towards Social Justice* in this book series, Baillie and Catalano (2009)). We are never suggesting we have worked out the right way to be and to act, but by introducing you to the mess and the complexity, we hope to show you that even amidst this chaos, it is possible to move towards what Deleuze and Guattari (1987) call 'cosmos.' What has any of this got to do with students and, you may ask, if we are we talking to engineers or students when we raise these questions? The answer is 'both.' We have framed the book around the idea that engineers taking part in development projects need to take action and prepare before they endeavour to enter this arena. It is just as valid for practising engineers who are currently working in industry as for engineering students. Many of the lessons learnt from the factory and the classroom may be transferable but not all of them are. Hence, we have decided to focus the book for the reader who is at any stage of their career as an engineer and who decides that he/she just doesn't know enough about working with communities different from her own. He/She can read some stories of engineers who have made that journey and, hopefully, learn from the many mistakes made along the way.

One of the authors of this book (Caroline) has worked with reduced environmental impact fibre reinforced plastics made from waste, for some time. These are materials like fibreglass, where the fibre makes the plastic stronger and gives it better properties. To reduce the environmental impact the 'matrix' is made from waste plastic and the reinforcing fibre is natural, sustainable, also waste. In different countries the potential natural fibres vary according to growing conditions. These can also be waste, e.g., waste from sugar cane in Brazil, flax stems after oil extraction in Germany, hemp stems in Canada or rice husk in China. When visiting Cairo a few years ago, I (Caroline) saw how the local family run rubbish pickers had purchased small processing machines called extruders and were using these to recycle the plastic themselves and sell an interim product, a pellet of plastic. This is easier to move around and much more valuable a product than squashed up 'bales' of waste plastic. Waste for Life (WFL) emerged out of the possibility of sharing my knowledge about composite materials with individuals, families and cooperatives who could merge this knowledge with existing businesses and make more money for themselves and their families.

In this book, we will follow the story of two different Waste for Life scenarios. Waste for Life (WFL) is a loosely joined network of scientists, engineers, educators, designers, architects, cooperatives, and artists as well as students of different disciplines and countries. WFL works at the intersection of waste and poverty using low-threshold/low-cost technologies to add value to resources that are commonly considered harmful or of minimal worth, but they are often the source of livelihood for society's poorest members. We will introduce you to two of the contexts within which WFL works currently – Maseru, Lesotho in Southern Africa and Buenos Aires, Argentina in South America. By juxtaposing the two contexts throughout this text, we hope that you will learn to identify the critical aspects that we focus on in each case. Though applied differently, the essence of what we are trying to do and why we doing it will be the same.

At this point, it is critical to note that the way we went about forming our partnerships for Waste for Life, is not the only way. There is no one size fits all. In some cases, NGOs may identify needs and reach out to engineering organisations such as *Engineers without Borders* in Australia or *Architecture for Humanity* and *Engineers Against Poverty* in the UK and make them aware of the projects that need support. Students might then get involved with these projects. However, the NGOs, EWB and EAP do not understand the complexities of the situations until they have worked with the local communities and seen the issues from many different perspectives. Even if the NGO is local, it can be biased, funded by problematic governments, industry or individuals who impose restrictions on their funding to further their own goals. Ideally, the community themselves should have some mechanism whereby they can directly contact an organisation that is willing to connect them with a network of experienced engineers who they can then educate about their community and its needs. In our case, I had decided that I no longer wanted to give my services and knowledge to those who could well afford it, or to companies who would only use it to profit themselves or their shareholders. I wanted to go back to the idea of engineering as service to humanity. Despite recent increasing awareness about the problems of 'engineering to help', it is clearly possible, given the right preparation and understanding of some key principles, to shift your 'client' from one group to another, from powerful and affluent to marginalised and vulnerable. However, it needs to be done well, collaboratively, be client directed and it needs to be done with a huge amount of awareness.

The book is presented as a series of chapters which do not need to be read linearly cover-to-cover, but may be dipped into as a particular focus area becomes relevant. WFL Lesotho and BsAs scenarios will be used to explore different aspects of our work. Chapter 2 begins by demonstrating how we entered into the community in Lesotho, working in a participatory way to identify possible avenues of collaboration. Chapter 3 goes into more technical detail to assess the feasibility once the initial need is ascertained.

Of course, we must explore the historical, political and socio-economic contexts in which we find ourselves. Chapter 4 lays out our approach to mapping the territory in Buenos Aires. We cannot start to work in an area before beginning to think through the potential impacts on all of the stakeholders. So we need to identify who those stakeholders are, how they can be involved, and how they might be affected. We need to decide how far we can go – clearly we cannot speak to everyone in the city or the country, so how do we know that what we are doing ensures that every group has been heard? We call this 'hearing all voices.' But even 'hearing' may not be enough because, remembering the warnings of Gramsci and Fleck, our interpretations of what we hear will always be translations from the original, moulded by our habits of thoughts, and that in those translations, something may be lost or ignored or suppressed that may need to be recaptured.

Chapter 5 focuses on one key stakeholder group in BsAs – the local government – in relation to another – the 'cartonero' or waste picker cooperatives. Chapter 6 takes this further by noting the complexity of the relations with and between cooperatives and the political complications of the BsAs context.

Once our partners have decided what they want to do – we need to work alongside them to ensure that it can, in fact, be achieved in an economically, environmentally, and socially sustainable way, long after our own participation has ceased (Chapter 7). Only after all of these conditions have been taken into consideration, can we hope to implement the project, which must be continually monitored against the community's measures of success. And after all that, we will still have questions and concerns. By raising these, we hope to help others learn from the lessons we learned. The most important thing to do both during and after this kind of work is to share our reflections, both good and bad, in an honest and humble way. We must be self-critical so that those who come after us can improve on what we have done. If we have such big egos that we need to be the only ones to ever do anything right, we should not start this kind of work as we will never learn from our mistakes and many people will suffer. In Chapter 8, we look explicitly at the role of students in development projects and the sorts of factors that need to be considered before bringing students into the program. We then summarise our thoughts and consider future actions (Chapter 9). Finally, we offer the reader some further reading to take with you a lifelong urge to know more and to constantly learn and question and develop your ability to make the world a better place, for you having lived in it.

The layout of this book is not meant to represent an order, timeline, or sequencing. At any given moment, we could be working on all of the issues in all of the chapters and some unexpected bit of information could send us scurrying back to our drawing boards for days to rethink everything before making another move. We cannot possibly represent the entirety of the complexity of entering into other peoples' worlds as change agents, nor ever hope to understand those worlds from all perspectives, but we must enter the maze and try as best we can to see and learn through the eyes of others.

It's best to approach this text as an instance of problem-based learning, which is how we approached our work in Lesotho and Argentina. This means that you'll be thinking along with us about development – its practices, benefits, and consequences – in a simulated environment.

Problem-based learning usually means that you are faced with a task and you use different resources to try to understand this task. You might, for example, use books, the Internet or lectures to give you the necessary preparatory knowledge. In a real development project, you would also need to talk to the local people, locate government documents, understand official policies, local power and social structures, etc. This means that you will be presented with the questions that we were faced with, and at times, you will realise that you do not have enough information to answer the questions we pose (just like us at that stage in the project). And you will need to think about what you might need to know in order to move forward. You will begin to see how we learned about our own problems through problem-based learning.

In each chapter, we will provide a series of questions and exercises, which might be used for independent study or in a class to explore the issues raised in the case studies.

CHAPTER 2

Assessing the Need in Lesotho

This chapter explores the environmental and socio-economic potential of a Natural Fibre Composite (NFC) business run by local cooperatives in Lesotho facilitated by Waste for Life. We selected to work in Lesotho because the key researcher in our laboratory at Queens in Canada was from there (Thimothy Thamae), and he suggested that the materials and process would be very beneficial for his country. Before embarking on such a project we knew we had to do much background work. We received funding to go ahead with these studies and began the exploratory phase of WFL Lesotho in 2005. The challenges of initiating such a project are considered here, in the light of an analysis of the economic and social contexts of the past decades and how they affected development projects similar to the ones we seek to initiate. The lessons of past efforts form the basis of the 'needs analysis' approach used in this study.

The Kingdom of Lesotho is a politically independent country, but it is a physical enclave of South Africa. It is about the size of the state of Maryland in the U.S., and its boundaries have remained constant since Britain annexed it as a protectorate in 1868 (the country gained independence in 1966). Lesotho's population is in decline, and has decreased from a high of 2.2 million in 2002 to 1.8 million in 2005. The United Nations projects that the death rate will continue to exceed the birth rate for the next forty-five years. The CIA World Factbook (2009) cautions laconically that 'estimates for this country explicitly take into account the effects of excess mortality due to AIDS; this can result in lower life expectancy, higher infant mortality and death rates, lower population and growth rates, and changes in the distribution of population by age and sex than would otherwise be expected.' Life expectancy is 34.4 years; 45% percent of the population is unemployed; 58% of the population lives below the poverty line; 35% of active male wage earners work in South African mines, the country being a labor reserve for South Africa. 86% of resident Basotho rely on some sort of subsistence agriculture to live, but in many areas the soil is exhausted, eroding and subject to recurring draught, which the recent report of the UN's Intergovernmental Panel on Climate Change (IPCC) predicts will only increase in regularity and severity; 60% of women ages 15-49 live with HIV/AIDS. Most progress made in human development and poverty over the past ten years is being rapidly reversed by the severity of the pandemic and, we now learn, global warming.

The first problem we imagined that WFL could help address in Lesotho is the management of waste plastic since Lesotho's urban centres have no well-established waste programs (Chapeyama, O., 2004; Mvuma, K., 2002). The lack of such programs results in much of the waste being dumped and burned in open spaces, a practice that poses serious health problems, especially in Maseru, Lesotho's main urban centre and a hub of industrial and commercial activities (Chapeyama, O., 2004). Mvuma, K. (2002) showed that plastic waste comprised a large share of waste generated in

Maseru. Lesotho, like other developing countries, generally resorts to collection and disposal of waste when faced with large, unmanageable quantities, rarely paying attention to other more sustainable waste management options.

Other economic opportunities using local resources present in this country include the untapped potential of agricultural residues as a profitable industry for Lesotho and the possible job creation through waste collection. Agriculture is the key sector in Lesotho, which is mostly a subsistence economy. Major crops include corn, wheat, sorghum, beans and peas. In most cases, these crops are mainly grown for the grain, and the remaining residues are subjected to uncontrolled grazing. These residues are potential fillers for the composites. Providing alternative uses for agricultural residues may not only increase the farmers' income; it may also boost agricultural production and multiply job opportunities in this sector.

The idea of initiating this kind of development project in Lesotho is not new. Millions of dollars have been spent in this country for decades in the form of aid from western governments and organisations. Despite good intentions, the results of such initiatives have been generally unsuccessful. In an effort to help Lesotho achieve some economic independence from a then apartheid South Africa in the 1970s and 1980s, many countries and organisations showed an extraordinary interest in assisting this country. In the period between 1975 and 1984 alone, Lesotho was receiving development assistance from 27 different countries and 72 non- and quasi-governmental organisations, leading Ferguson, J. (1990, p. 7) to ask, "What is this massive internationalist intervention, aimed at a country that surely does not appear to be of especially great economic or strategic importance?" Most of this "disproportionate volume of aid…given in astonishingly generous terms" (Ferguson, J., 1990, p. 7) hardly succeeded despite millions of dollars being spent. While this reality may lend itself to various interpretations, Ferguson identified one of the core reasons as the failure of Western donors to understand the way in which the local people viewed the projects or the commodities Westerners thought could be marketed in these developments.

One important example of these 'misunderstandings' involves the issue of land-use and land ownership. All rural land is owned and managed by local Lesotho communities (Leduka, 2006). Ignoring this fact is a very significant oversight, and many land development projects singled out a few people within a community and, in agreement with a chief, gave parcels of group land to these people, affectively excluding the rest of the community. Consequently, almost all of these projects failed because disenfranchised community members found ways to reclaim what they viewed as their legitimately owned land (Ferguson, J., 1990). They lost interest in government funded schemes that did not consult them. The 'supposed beneficiaries and their socio-political organisations were excluded from the policy formulation process' says Makoa, F. (1999, p. 47), who goes on to tell us that 'no serious studies were made regarding the role of the land tenure system in development and how people viewed it' (Makoa, F., 1999, p. 56). These developments are summarized (2006) in an earlier book in this series, *Engineers within a Local and Global Society*.

WFL attempts to work in another way, to conduct a needs analysis before any development is done and to work in a participatory way with local residents, the ultimate project beneficiaries. The study may not be extensive enough, but it is meant to give some insight into the realities of the local environment and what may expected from a successful project. The study investigates the present resources that could be used for making natural fibre composites in Lesotho and presents the results of interviews with the likely stakeholders in such a project, including farmers, householders, and members of cooperatives.

2.1   METHODOLOGY

Our first action was to survey local communities members to determine if there was any viability at all in a WFL collaboration to create NFCs or natural fibre composites from waste in Lesotho. We also needed to understand the scale of agricultural fibre sources, availability of waste plastics, and potential products, and we formulated some basic questions to find the answers (Table 2.1). Maseru district (one of the ten districts of Lesotho) was selected as the main focus for the project due to its suitability for producing both the waste plastic and fibre needed for the composites. Two visits were made to Lesotho, in 2005 and 2006. During the first visit, data was collected in two parts. The first part was obtained from published documents found in various institutions in Maseru city as shown in Table 2.2. The second part of the data was obtained from personal interviews with representatives of relevant institutions (Table 2.3). All interviews were conducted in Sesotho by our key researcher on the project, Thimothy Thamae.

It was clear from these initial interviews that WFL technology was welcome by many different groups. The question, then, was who should we work with? Plastics companies and local government agencies were quite interested in our project, but the most interesting group to us was the Cooperative College. It had been set up by the government, but the President felt that they were not being supported to the extent that they might be. The college provides training and technical equipment for many groups including Maseru Aloe, an umbrella cooperative whose members come to the college to process Agave gel from local Agave plants (see Baillie, C. (2006)). Further investigation and needs assessment was planned for this group.

The second action was to investigate the perceptions and views of relevant organisations, communities, cooperative members and individuals concerning the project. Short questionnaires were prepared for farmers, householders and members of cooperatives (Figures 2.1–2.4). The data were collected in the second trip in 2006 using these questionnaires as a basis for interviews and workshops (Tables 2.4, 2.5). Again, all meetings were held in Sesothu. The targeted groups were farmers, cooperatives and householders, who could work together from the supply to the end-use chain in the proposed industry. Telephone interviews of those already involved and interested partners were made as a follow-up to the interviews.

To find out from the local villagers what they thought the composite materials produced from waste could be made into, they were shown a piece of previously produced Agave/plastic bag composite and asked to consider possible applications. They often stated floor or roof tiles as shown below. Questions related to building materials for their houses were asked as follow-ups to determine their specific interest if, indeed, the product were to be used as tiles.

Table 2.1: Areas of focus.

Area	Questions	Objectives
Present sources of natural fibres	What are the present sources of natural fibres in Lesotho and how much of them are there?	To investigate if there are suitable sources of fibre existing in the right amounts for NFCs industry
Waste plastic production	What kind of plastic and how much of it is produced in Lesotho?	To investigate if the recyclable plastic waste exists in enough quantities to sustain NFCs industry
Waste plastic recycling institutions and technology	What kind of related technology and institutions already exist? What role can these institutions and technology play in this industry?	To supply information relating to institutions that could work together to develop the industry, taking advantage of existing local technology

Table 2.2: Document analysis.

Type of information	Name of the document	Source organisation
Present sources of natural fibres and their costs	Lesotho agricultural situation report (2000/01-2001/2002)	Lesotho Bureau of Statistics
	Lesotho agricultural production survey of crops (2003-2004)	
	Names and numbers of Agave americana cooperatives in Lesotho	Lesotho Producers Cooperatives (LEPCOP)
Information about waste plastic	The cleanest town competition report (May, 2004)	Lesotho National Environmental Secretariat
	Lesotho plastic companies	Lesotho Ministry of Trade and Industry
	National environment youth corps, annual report (1998)	UNDP Lesotho
Information about use of present roofing materials	Lesotho environmental statistics	Lesotho Bureau of Statistics

Information needed	Institution/business	Position
Present Sources of natural fibres	Ministry of Agric: Crops section	Contact person
	Ministry of Agric: Crops section	Chief crop inspector
	Cooperatives College	Lecturer
	Mahloenyeng trading	Contact person
Information about local weaving	Meliehe Weavers	Contact person
	Matela Weavers	Contact person
	Seithati Weavers	Contact person
	Thabong Weavers	Contact person
	Itjareng vocational training centre	Contact person
Information about plastic and plastic waste	MU Plastics	Managing director
	T'sosane plastic waste scavengers	Representative
	MOS plastic Manufacturers	Contact person
	Pioneer Plastics	Contact person
		Managing director
	Appropriate Technology Services	Chief engineer
	National Environmental Secretariat	Contact person
	Masianokeng Environmental centre	Contact person

Table 2.3: Interviews undertaken during first visit.

2.2   RESULTS

The data collected from the second visit are summarised in Tables 2.6 and 2.7 below.

In a 2001 demographic survey in Lesotho, it was shown that nearly two thirds of the houses used corrugated iron as roofs (Lesotho Bureau of Statistics, 2001). The figure rises to 90% in the urban areas and drops to 57% in the rural areas where grass or straw are the alternative traditional roofing material. These alternative materials (grass and straw) are becoming scarce due to poor rangelands management (Marake et al., 1998). Lesotho's hot summer months and exceptionally cold

Name and location of the farm...
1. What crops do you normally grow?
2. On average, how much yield do you produce for each crop a year?
3. Do you do subsistence farming, cash farming or both?
4.   What do you do with your crop stover or straw? Give it to animals, burn it, take it back to the soil, other (explain)
5. Would you sell your stover/straw if opportunity arises? Yes, No, Why?
6. If you were to sell your stover/straw, what factors would you consider before putting a price?
7. Are you aware of any other alternative uses for this stover/straw? If yes, what are they?
8. Do you think Agave Americana is something you can cultivate for commercial purposes?

Figure 2.1: Questionnaire for farmers.

Name and location of the village...
1. How many houses do you have?
2. How many people live in your house?
3. If you had enough money, what roofing would you choose and why?
4. When you look at this sample, what products do you think it can be used for?
5. Giving priority, which of those products do you think most people in your neighborhood really need?
6. What kind of roofing your houses have: tiles, corrugated iron, thatch or other? Explain 'other'.
7. If you used a thatch for roofing, did you buy it or did you get it from public areas for free?
8. Are you satisfied with the temperature levels in your house in both summer and winter? Why?
9. Do you think roofing has anything to do with your house temperatures? Why?
10. Do you consider the corrugated iron roofing as 'cheap, moderate or expensive'?

Figure 2.2: Questionnaire for householders.

winters make the use of insulating ceiling panels below poorly insulating iron roofs a requirement. However, only a few people can afford to purchase these insulating ceiling panels. Our studies indicated that cheap ceiling panels were among the most desirable products to be made from natural fibre composites in the country.

According to Lesotho Bureau of Statistics (2002), iron roofed houses makes up 62% of the all the houses in Lesotho, followed by thatch/straw, 35% and roof tiles, 2.1%. In our study, we achieved similar results with 67% of the houses belonging to the interviewed population roofed with corrugated iron, followed by 22% thatch and 10% roof tiles. The higher than national average percentage of roof tiles is likely due to the fact that a significant number of the interviewed population were from urban and semi-urban areas where roof tiles are more common. As stated before, the use of thatch, a traditional roofing material, which is perceived locally as a good insulator is declining due to poor management of resources. It was traditionally harvested for free from community owned rangelands, but as found in this study, 40% of the respondents now bought it.

Name and location of the group...

Source of Agave Americana

1. Do you grow your own Agave Americana or do you harvest the existing one from public areas. If you harvest the existing one, do you do anything to ensure its sustainability? Explain.
2. From your sources of Agave Americana, do you think there is enough Agave Americana for products you want to make?
3. Do you have to get legal permission to grow or extract Agave Americana leaves? If so how do you get permission?

Processing

4. How do you cut the leaves from the whole plant?
5. How do you open the leaves to get the product you want out of them?
6. What do you do with the Agave Americana fibre after getting the product you were looking for?

Products

7. What products are you making out of Agave Americana leaves?
8. What products do you still hope to make out of Agave Americana leaves?

Challenges

9. What are some of the problems you get in working with Agave Americana plants?

Figure 2.3: Questionnaire for spokesperson of co-ops.

Name and location of the group...

Demographics

1. Male or female?
2. How old are you?
3. Did you have any schooling, if you did, what class did you do?
4. Do you have children, if so, how many?

Income and Its Sources

5. How much income per month do you make out of the Agave Americana products sales?
6. Do you have other occupations? What are they?

Possible Applications

7. When you look at this sample, what products do you think it can be used for?

Production Process

8. What do you think is the easiest method to remove fibre from the leaves?

Figure 2.4: Questionnaire for members of co-ops.

Figure 2.5: Thatched roof in Lesotho.

Figure 2.6: Thatched roofs in Lesotho.

Figure 2.7: Tin roof in Lesotho.

Figure 2.8: Tin roofs in Lesotho.

Figure 2.9: Tin and thatched roofs in Lesotho.

Figure 2.10: Villager outside her thatched roof home in Lesotho.

Table 2.4: Interviews with local groups in Maseru district.

Area/group	Location in Maseru district	Interviewees	Number of interviews
Rural areas	3 villages at Roma area	Householders	60
		Farmers	60
Semi-urban and urban areas	3 semi-urban dwellings in Roma and 1 locality in Maseru	Householders	54
		Farmers	10
Maseru Aloe groups	Maseru	Representatives of groups	11
		Members of groups	25
Total number of interviews			220

Table 2.5: Workshops held with members of Maseru Aloe and other co-ops.

Location in Maseru and Roma	Group name	No of attendees
Semphetenyane	Boiketlong Multipurpose	11
Motimposo	Ithabeleng Multipurpose	20
Ha Tsosane	Boiketlong Multipurpose	9
Roma		10
Stadium area	Hlabollanang Multipurpose	9
Maseru down town	Lechabile Multipurpose	1
Total number of attendees		60

Whereas the householders mainly lived in corrugated iron roofed houses, the majority of them (45%) would rather be living in tile roofed houses since tiles have a 'good' appearance. Good appearance, associated mostly with tiles, was therefore the most cited reason (20%) for choice of roofing. The second and the third most cited reasons for choice of roofing were ability to insulate (18%) and durability (15%), attributes which were mostly associated with thatch and iron roofs, respectively. In winter, many iron roofed houses experience water condensation on the roofs due to steam from cooking, resulting in water dripping known locally as 'rothe.' This phenomenon does not occur in iron-roofed houses with insulating ceilings because they do not cool enough to permit condensation on their surfaces.

Demonstrating the serious lack of insulation in most houses, 85% of the interviewees were not satisfied with the temperatures in their houses, which were considered too hot in summer and too cold in winter. About 93% of the householders associated roofing with these extreme temperatures. The

Table 2.6: Farmer Responses.

Question	Answer	%	Question	Answer	%	Question	Answer	%
What crops do farmers grow?	Maize	39	What is farming done for?	Subsistence	67	What is the straw presently used for?	Animals	88
	Sorghum	28		Cash	7		Back to soil	5
	Wheat	3		Both	26		Burn it	7
	Legumes	24						
	Vegetables	7						
Would famers sell their straw?	Yes	76	Why would some not sell their straw?	Animals	82	Are farmers aware of alternative uses for straw?	Yes	51
	No	24		Other	18		No	49
What alternative uses of straw are farmers aware of?	Heating	31	Are they willing to grow agave for commercial purposes?	Yes	90			
	Animal food	25		No	10			
	Manure	39						
	Handicrafts	6						

Table 2.7: Householder responses. (*Continues.*)

Question	Answer	%	Question	Answer	%	Question	Answer	%
What are the reasons for preferring a particular kind of roofing?	Cheap	8	What are possible uses of the composites?	Roofs	13	Source of thatch?	Bought	40
	Wind resistant	14		Ceiling	16		Free	60
	No water dripping	7		Floor tiles	30		Thatch	32
	Needs no expertise	8		Wall tiles	8	Preference of roofing type?	Tiles	45
	Durable	15		Furniture	13		Iron	22
	Insulating	18		Face board	5	Are householders satisfied with temperatures levels in their houses?	Yes	15
	Good appearance	20		Other	14		No	85

Table 2.7: (*Continued.*) Householder responses.

Question	Response	Value
	Good for cooking	4
	Noise resistant	5
	Fire proof	2
Does roofing affect in-house temperature?	Yes	93
	No	7
What roofing is used in each of the buildings householders have?	Tiles	10
	Corrugated iron	67
	Thatch	22
How do householders view the cost of iron roofs?	Cheap	16
	Moderate	29
	Expensive	56
What are the reasons for not being satisfied with temperatures levels in the house?	Summer fine, Winter too cold	39
	Winter fine, Summer too hot	5
	Summer too hot, Winter too cold	57

main culprits for this situation were the iron roofs. Nevertheless, the majority of people eventually use iron because it is far cheaper than tiles, and unlike both tiles and thatch, it needs little expertise in the roofing process. When they were given small flat samples (about $10 \times 10 \times 0.2$ cm) of composites made from agave, corn straw, and waste plastic and asked to suggest uses, householders gave a range of them. The most cited product was floor tiling. A range of products people think about when they see the composites is the first step in getting to know what products are really desired.

A few lessons can be learned from the responses. Introducing NFCs ceiling panels would help moderate room temperatures in iron roofed houses without a need to change the present roofs. However, for this to be successful, the ceilings have to be extraordinarily cheap. Otherwise, most people could not afford them as they cannot now afford the existing ceiling panels in the marketplace. They also have to be very durable and need little expertise to mount to the roof. Most importantly, they should require no change in the present roofing system to accommodate them. Since most people preferred products that 'looked good,' appearance of the final products must not be overlooked at the expense of providing them cheaply.

Success in making and using NFCs would depend on the interest of and impact on farmers who produce straw or Agave americana as well as impact on the land. As expected, the majority of farmers (39%) grow corn, followed by sorghum (28%). These crops are grown mainly for subsistence (67%). Only 33% of the farmers sell their products and just 7% grow their crops exclusively for cash. If given the opportunity 76% of the farmers would sell their straw. This is a significant percentage and may have positive implications for the NFCs industry. The farmers who would not sell their straw cited the need to feed animals as the reason in 88% of cases. Almost half of the respondents were not aware of any alternative uses of straw other than how it was already being used. Those farmers who were aware of alternative straw uses, were mostly aware of its use as manure for a healthy soil (39%), only 5% of the farmers used straw as manure. Using straw as manure could mean decomposing it several months in soil well before the planting season. This is a helpful process which may prove costly for farmers since it means double –tilling each year. A huge majority of farmers (90%) would be willing to cultivate Agave americana for commercial purposes.

Regarding the manufacture of the composites, whilst several local companies expressed interest in producing NFCs, it was deemed better to focus on cooperatives, which would benefit a broader spectrum of people. More than a 100 different cooperatives were identified in all ten districts of Lesotho. One such cooperative was Maseru Aloe whose members were extensively interviewed. This is an umbrella organisation of around 14 different cooperatives based in Maseru city. Under the direction and assistance of the Lesotho government department of cooperatives, this body takes juice from Agave americana leaves to make mainly skin gel and creams. The products are sold locally and to neighbouring South Africa. Results of these interviews are shown in Table 2.8. These cooperatives consisted of mainly women (88%). The age average was 53 years. An average member had at least completed the last year of elementary school and had 4 children. Most of the respondents took farming as an alternative occupation (30%), 19% made and sold handicrafts, and 22% had no

Table 2.8: Demographics of members of cooperatives interviewed.

Question	Answer	Response	Question	Answer	Response (%)
Gender	Percentage of males	12 %	Other occupation	Farmer	30
	Percentage of females	88 %		Tailor	15
	Mean age	53		Nurse	7
Other demographics	Mean class	7		Make and sell handicrafts	19
	Mean number of children	4		Sell wood	7
				None	22

alternative occupation. The age distribution shows most members falling between the age of 50 and 60 (Figure 2.11). Ages ranged from 27 to 72.

During interviews and workshops, all representatives and workers of the cooperatives agreed that the plant fibre waste plastic composites would be an excellent additional product to create within their groups. They have access to waste plastic, straw and Agave americana fibre, and they can learn the manufacturing techniques and house the necessary equipment at the cooperatives training centre where the Agave americana gel and creams are presently made.

One of the major blocks to implementing a composite manufacturing facility in Lesotho was deemed to be the high cost of equipment. Therefore, low cost manufacturing would be a priority if composite projects were to succeed. Seeking low-cost alternatives, the Department of Mechanical Engineering at Queen's University worked collaboratively with the University of Napoli, Italy, Lerotholi Polytechnic in Lesotho, Department of Cooperatives, and Maseru Aloe Cooperatives in Lesotho, to design a compression moulder or 'hot press' prototype shown in Figure 2.12. This basic machine can be used to make ceiling tiles and other flat panels using waste plastic. The hot press is designed to be manufactured from inexpensive materials that can be purchased locally or in neighbouring South Africa. Like any other hot press, the materials to be made are placed in a mould between two heated platens and compressed to desired shapes. Lerothli Polytechnic expressed an interest in developing a business side to their educational profile and manufacture the hotpresses on demand.

This study balances the need to provide technical assistance in building up NFC industry in Lesotho with the need to co-create solutions that are relevant and directly address the needs of the local people. It is argued that the massive failure of previous development projects in this

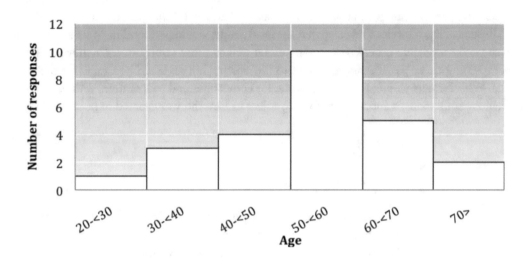

Figure 2.11: Members of cooperatives according to different age ranges.

Figure 2.12: A hot press designed by Darko Matovic at Queens University.

country were partly due to lack of this or similar forms of consultation, which we call needs analysis, understanding and providing what people need rather than what developers think they need.

Results of the interviews revealed that the local people would prefer ceiling panels as a product as these would moderate otherwise extreme winter and summer temperatures aggravated by iron roofs. Providing cheap panels, easy to mount to the existing houses could be a relief for the house-holders. With farmers willing to sell their straw for composites and cooperatives willing to take up the challenge of making composites, an NFC project in Lesotho was deemed to be a possibility.

EXERCISE

The Failures of development

Class discussion: How do you view the WFL initiative in relation to the development failures of the past in Lesotho? The discussion can centre around some of the key questions that we have asked ourselves. One pos-sibility might be to invite students from development studies or sociology into the classroom, either in person or virtually via video link up or skype to help frame these ideas in a 'post development' context.

1. Is this a technology driven development? Is this a solution looking for a problem? Is that ok? Or should the needs come first?

2. Is this an example of what Cowan and Shenton (1996) would call trusteeship, i.e., citizens in the North thinking that they know bet-ter/more and can therefore 'help' citizens in the South?

3. Who really benefits? What is the role of the International University players?

4. Is the needs analysis with the local community patronising or partic-ipatory? Should we avoid the word 'needs'?

5. How do we stay in touch with our partners on the ground once we are back home?

CHAPTER 3

Feasibility of WFL Lesotho

Once the potential and basic outline of the program is ascertained, the next stage is to study the feasibility in detail so that the risks are known beforehand and can be measured against any perceived benefits. In this chapter, we look more closely at the source of fibre and waste plastic in Lesotho, as well as the potential product, ceiling tiles, to assess feasibility.

3.1 SOURCES OF NATURAL FIBRES

Major crops in Lesotho include corn, sorghum, wheat, beans and peas (Lesotho Bureau of Statistics, 2002). There are no fibre crops in this country. Corn is the most widely cultivated crop because its grain is a local staple food. The next most widely cultivated crop is sorghum.

On average, the area of land used to produce corn has been about three times as much as that used for producing sorghum per annum over the past two decades (Figure 3.1). Corn grain production is roughly four times more than that of sorghum per annum over the same period. It contributes 67% of all grain production per annum (Table 3.1, Figure 3.2). There is an upward trend of corn production even though the area devoted for it declines slightly over this period (Figure 3.3). High fluctuations in area planted and grain harvest can be expected in a country where crop production is heavily dependent on weather.

The crops discussed above are not sources of strong fibres. The Agave americana plant was identified as the only source of strong fibres in the country. Although this plant grows in abundance all over the country, there are no records concerning how much of it exists. One of its advantages is that it grows in all four topographical regions of the country: Lowlands, Highlands, Foothills and Senqu Valley. It does not need too much care; once it is planted, it continually renews itself as seedlings keep growing. Its disadvantage is that it takes a long time to mature, hence its common name, century plant.

Since it is not presently cultivated as a plant, Agave americana plants are not found concentrated in significant amounts in one place. Rather, they form small clusters scattered all over the country, especially in the countryside where they were traditional used for fencing. It is presently used by a number of cooperatives in the production of Agave skin gels which results in fibre as a waste product. Plans to make use of large amounts of this plant in the composite industry would need careful management. This could include efforts to cultivate it as a crop. Otherwise, the existing plants could be overharvested. The scattered clusters of this plant could also create a challenge in collection.

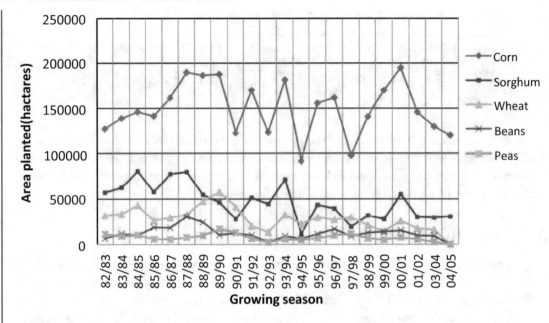

Figure 3.1: Area planted for major crops in Lesotho over time (Lesotho Bureau of Statistics, 2002).

Table 3.1: Average production of grain and straw from major crops in Lesotho Bureau of Statistics (2002).

Parameter	Corn	Sorghum	Wheat	Beans	Peas	Total
Average grain production per annum (metric tonnes)	117848.3	31252.5	20236.1	5467.2	2531.4	177335.5
Average straw production in per annum (metric tonnes)	235696.6	62505.0	40472.2	10934.5	5062.8	354671.1
Percentage	66.5	17.6	11.4	3.1	1.4	100.0

3.2   PRODUCTION AND RECYCLING OF WASTE PLASTIC

The amount of waste plastic produced in Lesotho is not well known. Only one report, a PhD thesis by Mvuma, K. (2002) gave an idea of how much plastic may exist at that time. The study focused on two areas, Maseru and Maputsoe, which are the two main industrial and commercial hubs in Lesotho. Table 3.2 shows the contributions of each sector in the total annual waste plastic produced. The commercial establishments (e.g., grocery shops) contributed the most waste plastic, even more than industries. Waste plastic shared a large percentage of solid waste produced (36%) that is 48342.9

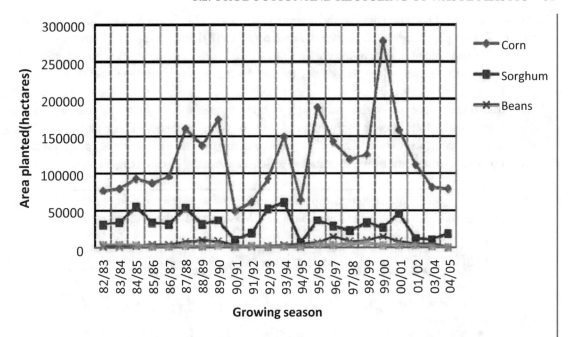

Figure 3.2: Grain production of the major crops in Lesotho over time (Lesotho Bureau of Statistics, 2002).

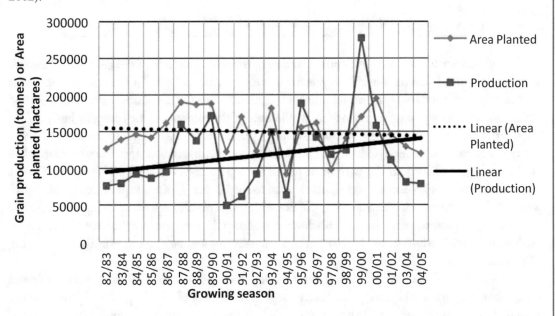

Figure 3.3: Trends in the area planted and production of corn grain in Lesotho Bureau of Statistics (2002).

tonnes per annum. This is surpassed only by waste paper, 54%. This figure does not likely reflect the amount of waste plastic produced all over the country in more than 10 other small towns.

Only one waste plastic recycling company of Chinese origin (MU plastics) was identified in Maseru. It recycled mainly plastic bags from the neighbouring textile companies and commercial establishments. The plastic is sorted, cleaned and shredded before being passed through an extruder. The extruded material is pelletized and mixed with virgin plastic (bought from South Africa) to make mainly plastic bags though extrusion and blow molding processes. These plastics are later resold mainly to the grocery stores. The main challenge in this process was mentioned as the presence of small rocks and metals in the waste material, which occasionally blocked the extruder.

Table 3.2: Annual composition and quantity of waste generation per category and source in surveyed areas of Maseru and Maputsoe, Lesotho (Tonnes/Annum) (Mvuma, K., 2002).

Category	Value (t/annum)
Low income	72.38
Middle income	9.94
High income	83.12
Industries	29.54
Commercial establishments	48147.95
Total	48342.9
Percentage of plastic in total solid waste	36%

Another waste company (Welcome Transport), which originated in 1980 and had 70 employees was identified. It did not directly recycle the waste plastic but collected it (and many other kinds of waste), sorted it and sold small amount to MU plastics. The rest of the plastic was sold to recycling companies in Johannesburg, South Africa. At the time of the interviews, the company had plans to recycle the plastic on the site. It was also installing a plastic shredding facility to make transport of bulky plastic easier in order to reduce transportation costs. It worked directly with an estimated 450 waste scavengers who supplied it with waste plastic for a price of M 0.5/kg (1 Maloti = 13 US cents at the time of writing). Welcome Transport claimed to send 8 tonnes of waste plastic to South Africa a month. It also burns a significant amount of unusable waste plastic each day. The biggest challenge to the company was a competition with what the company interviewee described as 'shady' South African companies which came directly to the waste scavenges to buy plastic, effectively bypassing Welcome Transport.

One informal waste scavenger supplying waste plastic to Welcome Transport was interviewed on the dumping site. She claimed to make an average of M20 a week from the process. Her challenges included the constant burning of the waste plastic by some members of the public, the practice of which reduced her source of income. It is important that any future waste plastic projects do not

Figure 3.4: The main dumping site at Ha T'sosane in Maseru, Lesotho. People disposing and scavenging for waste can be seen as well as smoke from some of the waste.

sideline the obviously poor scavengers who are making ends meet with this job. Rather it would be important to make them part of the project. These relationships can be summarized in Figure 3.5.

3.3   A ROOF CEILING SCENARIO

Cheap roof ceiling could help provide highly needed insulation in both summer and winter for householders in Lesotho. For simplicity, the houses can be taken as having similar shape and size and as following the model most common in Lesotho, polata (Figure 3.6). We pose the question, how many houses and at what cost could the houses be insulated given the present production of fibre? Suppose we decide to estimate the number of houses N_h which can be insulated with thin composite panels made from waste plastic and milled straw fibre from the existing local crops. The panels are thin square prisms, with the base area A_c capable of insulating a roof area A_h, which is equivalent to floor area in the house. To better focus on what the fibres can do, we assume unlimited supply of available kind of waste plastic (M_m) but limited supply of fibre. Therefore,

$$N_h = \frac{n_c A_c}{A_h} \tag{3.1}$$

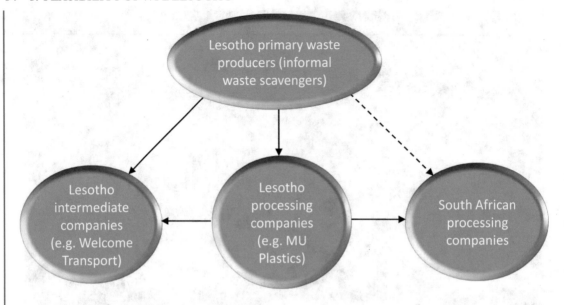

Figure 3.5: Relationships between waste plastic producers and users in Lesotho and South Africa.

where n_c is the total number of panels that can be made from available straw and unlimited waste plastic in the country every year. The number n_c is a function of the total mass of all the composite material M_{ct} that can be made from available straw and unlimited waste plastic each year and the mass of each composite ceiling panel M_c of base area A_c. Thus,

$$n_c = \frac{M_{ct}}{M_c} .$$

(3.2)

Assume that the whole straw and not only fibre is used for reinforcing the plastic. Then the total mass of all the composite material that can be produced each year M_{ct} would equal the total mass of straw available for use in composites each year M_s divided by its weight fraction m_f in the composite. This amount of available straw can be obtained from total grain production per year M_g by dividing it with its Harvest Index (HI) or grain to straw mass ratio of the crops in use (see below for more information on HI). M_s would also depend on the fraction of available straw to the total straw produced in the country each year, a_f (for various reasons, not all straw produced will be available for use in NFCs).

$$M_{ct} = \frac{M_s}{m_f} = \frac{M_g a_f}{HI m_f} .$$

(3.3)

The mass of each composite panel M_c is:

$$M_c = \rho_c V_c$$

(3.4)

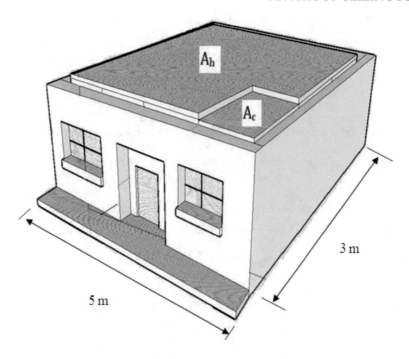

Figure 3.6: A typical corrugated iron roofed house popular in Lesotho usually referred to as 'polata' in local language. A_h is all the roof area between the walls of the house, it is equivalent to the floor area. This section of the roof is 'cut' to show the area A_c covered by a square ceiling below it.

where ρ_c is the density of the composite panel and V_c is its volume. If v_f, ρ_f, and ρ_m are volume fraction of straw in a composite, density of straw and density of polymer matrix in a composite, respectively, then, from rule of mixtures (Hull, D., 1990):

$$\rho_c = v_f \rho_f + (1 - v_f)\rho_m .$$ (3.5)

Thus:

$$M_c = V_c \left[v_f \rho_f + (1 - v_f)\rho_m \right] .$$ (3.6)

From (3.2), (3.3), and (3.6),

$$n_c = \frac{\dfrac{M_g a_f}{H I m_f}}{V_c (v_f \rho_f + (1 - v_f)\rho_m)} .$$ (3.7)

However, v_f can be expressed in terms of weight fraction of straw in the composite m_f using the relationship (Hull, D., 1990).

$$v_f = \frac{\dfrac{m_f}{\rho_f}}{\dfrac{m_f}{\rho_f} + \dfrac{1 - m_f}{\rho_m}}. \tag{3.8}$$

From (3.7) and (3.8), the number of panels that can be made from known grain harvest

$$n_c = \frac{M_g a_f}{V_c H I} \left(\frac{m_f \rho_m + \rho_f - \rho_f m_f}{m_f \rho_f \rho_m} \right). \tag{3.9}$$

If there is a need to assess the number of composite panels that can be made from a measured amount of available straw, Equation (3.9) can be simplified to

$$n_c = \frac{M_s}{V_c} \left(\frac{m_f \rho_m + \rho_f - \rho_f m_f}{m_f \rho_f \rho_m} \right). \tag{3.10}$$

From (3.1) and (3.9), the number of houses that can be insulated from known grain harvest per year

$$N_h = \frac{A_c M_g a_f}{V_c H I A_h} \left(\frac{m_f \rho_m + \rho_f - \rho_f m_f}{m_f \rho_f \rho_m} \right). \tag{3.11}$$

It may be necessary to estimate the cost of producing these panels and compare the price with that of conventional panels in the markets. The cost of producing each panel C_c

$$C_c = m_f M_c C_f + (1 - m_f) M_c C_m + C_p \tag{3.12}$$

where C_f, C_m and C_p are cost of straw per unit mass, cost of waste plastic per unit mass and cost of composite processing per panel, respectively. From (3.1) and (3.12), the total cost of making composites that would insulate a known number of houses C_{ct} would be

$$C_{ct} = \frac{A_h N_h}{A_c} \left(m_f M_c C_f + (1 - m_f) M_c C_m + C_p \right). \tag{3.13}$$

The following crude estimates and assumptions can be made to estimate the number of houses that can be insulated from available straw in Lesotho and the cost of producing composites to insulate them. For simplicity, straws from different crops are assumed to have insignificant differences in reinforcing capabilities. Consider an annual average grain production of all major crops in Lesotho of 1771335 M (Table 3.3). This figure can be used to calculate the amount of average annual straw produced in this country (crop production is normally given in grain rather than straw figures). To do this, the concept of HI or grain-to-straw mass ratio of the crops is used. As can be expected for natural materials, this property varies even within the same plant species. Prihar and Prihar (1991) had the following ranges of harvest indices: 0.48-0.53 for sorghum, 0.58-0.60 for corn and 0.38-0.47

for wheat. Hay, R. (1995) suggested that HIs of the most intensely cultivated grain crops generally range from 0.4 to 0.6. The crops in Lesotho fit this description due to their common use as sources of food worldwide. Therefore, it is assumed that the HIs for each of these crops range around the average of 0.5. With this HI, the annual average straw production M_s has been calculated.

The question of how much straw can be available for use in composites depends on many factors such as willingness of farmers to sell it (and their need for cattle feed) and ease of transporting this straw to the project site. Since estimating such factors would be a challenge, the middle number, 0.5 (50%) is arbitrarily taken as the fraction of available straw to the total straw produced in the country each year, a_f (which means half the straw produced in the country is assumed to be available for use in the composites). Of course, this fraction can be varied to note the differences in whatever parameter is being estimated.

From observation, there is a tendency to build a number of small one or two roomed houses rather than one big house with many rooms in Lesotho. The most common house is a small corrugated iron roofed house made from concrete bricks whose average floor area dimensions A_h can be estimated as 5m × 3m. Suppose a manufacturing site produces 60 cm× 60 cm× 0.6 cm ceiling panels (this is the size of the panels planned to be used in the envisaged project in Lesotho). Each panel would insulate a roof area A_c of 0.36 m^2 inside the house and have a volume V_c of 0.00216 m^3.

The densities of waste plastic and straw can be estimated as 0.95g/cm^3 (for instance, HDPE bags common in Lesotho) and 1.2g/cm^3 (common for natural fibres, although in straw, there is more than just natural fibre), respectively. The weight fraction of straw in the composite will be taken as 0.3. The cost of waste plastic, straw and processing are taken as 0.6 M/kg, 0.5 M/kg and 0.7M/panel, respectively. These crude estimates can be refined as more information becomes available. They are merely meant to give a reasonable idea of what the present size of raw materials in the country could achieve. The arbitrary numbers should not be viewed as facts. All the estimates are summarized in Table 3.3.

With these assumptions, the number of houses that can be tiled starting with the known annual production of grain of 177335.5 t in Lesotho can be estimated at 6.5 million (Figure 3.7). When unlimited waste plastic is assumed, there is unlimited number of houses that could be insulated when the fibre weight fraction used in the composites approaches zero (Figure 3.8). However, given a limited supply of straw, increasing its weight fraction in the composites rapidly reduces the number of houses that could be insulated.

From Equation (3.13), we can see that it would take just M29 for the raw materials to make produce ceiling panels enough to insulate a single house (see also Figure 3.9).

The results show that any NFCs project that can be initiated in the near future in Lesotho would depend on available straw from the commonly cultivated crops and competition for use as a food course for cattle. Agave americana fibre also has great potential to be used in composite in this country if accompanied by proper management which would include cultivating it in selected areas like other crops. The results above help in finding which crops are more likely to be targeted for

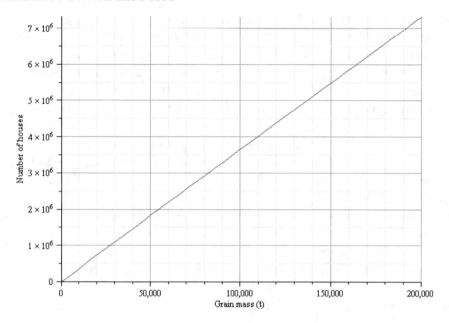

Figure 3.7: Number of houses that can be insulated from annual grain production in Lesotho (plotted from Equation (3.11) using estimates in Table 3.3 and assuming weight fraction of 0.3).

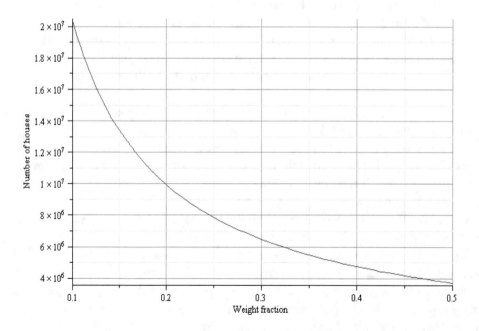

Figure 3.8: Effect of weight fraction of straw per panel on a number of houses that could be insulated given unlimited waste (plotted from Equation (3.11) using estimates in Table 3.3).

Table 3.3: Parameters for estimating the number and cost of roof ceiling and number of houses they can insulate in Lesotho.

Property	Symbol	Estimated value
Grain production per year	M_g	177335.5 t
Straw production per year	M_s	354671.1 t
Mass of available waste plastic	M_m	unlimited
Harvest index	HI	0.5
Roof area needing insulation	A_h	15 m^2
Roof area each ceiling panel could cover	A_c	0.36 m^2
Volume of each ceiling panel	V_c	0.00216 m^3
Density of straw	ρ_f	1.2 g/cm^3
Density of waste plastic	ρ_m	0.95 g/cm^3
Straw weight fraction	m_f	0.3
Fraction of available straw	a_f	0.5
Cost of straw	C_f,	0.5 M/kg
Cost of fibre	C_m	0.6 M/kg
Cost of processing	C_p	0.7 M/kg

use in the composites, both for a short term and long term plans. Nevertheless, they say little as to whether the crops are in adequate quantities for the proposed use in NFCs. Solving this problem is a challenge since there should be some reference against which to measure such adequacy. That would be the level of demand for the NFCs in the country which does not presently exist. Additionally, there would appear to be enough waste plastic to make the business a viability, but it would be important to include the scavengers in such a process as with our next case study in Buenos Aires.

In a fictional project supplying roof ceiling for a common form of houses in Lesotho, it was estimated that up to 6.5 million of such houses could be insulated at the raw material cost of nearly M30 for each house if there were no competition for the fibre source and no other negative impacts. Of course, then energy costs, repair, housing and time would have to be considered. These are discussed further in Chapter 7.

At this stage, WFL Lesotho was considered to be worthy of further funding and exploration. At the time of writing, funding has just been awarded from the United Nations Development Program Global Environment Facility to support Maseru Aloe in the development of the business. Thimothy Thamae has returned to Lesotho to start work at the University of Lesotho and will support the work from there together with an international team of volunteers as indicated below in Figure 3.10, as well as the University of Western Australia who has recently joined in with the work. Students and academics from Canada, Italy and Australia will continue to feed into the ongoing development of this project in Lesotho.

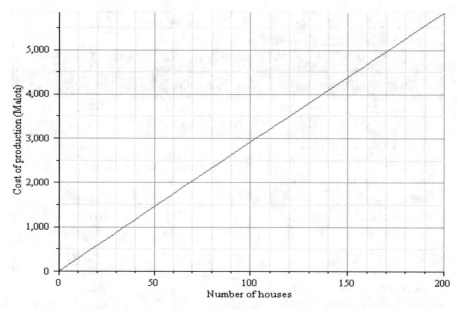

Figure 3.9: Cost of producing straw waste plastic composites as a function of a number of houses being insulated (plotted from Equation (3.13) using estimates in Table 3.3).

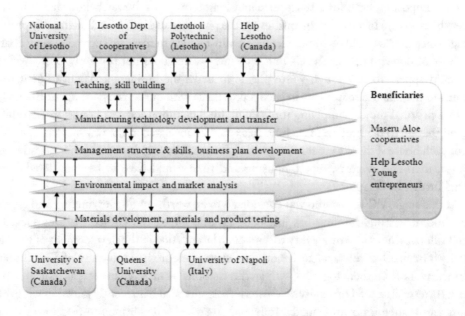

Figure 3.10: Partnership map: horizontal arrows represent 5 layers of project activities; vertical arrows indicate each institution's link with a particular project layer.

EXERCISE

Feasibility studies

Create a Feasibility Issues Table for WFL Lesotho as below. Make an assessment of each issue and decide where more information is required. Think of more categories we have not considered.

Issue	Comments
Agricultural sustainability: Is Agave sustainable as a product? Will straw compete with food source for cattle?	
Market: Who will buy the ceiling tiles? Who needs them but can also afford them? For those who can't, is there another source of funding to buy them? What about the implications of a South African market?	
Local stakeholders: How can we ensure that no-one suffers from WFL initiative, e.g., scavengers? Can they be brought on board?	
How will the start-up project be funded? What are the implications of development agencies or government being involved?	
What is the role of the international university players? How can we ensure that this is not just benefiting them?	

CHAPTER 4

Mapping the Territory in Buenos Aires

Waste for Life Buenos Aires began very differently than WFL Lesotho. We had no direct links to the country and, therefore, the initial needs assessment stage took much longer. After visiting Argentina and seeing the paper and plastic being collected by the waste pickers or 'cartoneros' in the streets of Buenos Aires, WFL wondered whether the technology might enable cooperatives to remain independent and economically autonomous by developing a manufacturing branch to their collecting and sorting work.

In this chapter and the subsequent chapters, we will present our interactions with the network that we encountered and some of our self-critique about this process and lessons learnt. We attempt to address some of the following questions:

- Can Waste for Life help cartonero cooperatives become more autonomous and self-sufficient?

- What materials would be used and what products made?

- Could the hotpress as designed for Lesotho work in Buenos Aires? Could it be made there and by whom?

- How could the cooperatives sustain the momentum after the life of the WFL project?

Two WFL team members visited Buenos Aires (BsAs) for six months in 2007 and again in 2008. During these trips, we explored possible work scenarios that included service learning projects for students. The narrative of what we did, who we did it with, and how we did it is fairly easy to tell. The questions, the heartache, and the uncertainties are not. We did not enter into this project lightly. We did not go down to Buenos Aires innocently hoping to help. We were constantly plagued by questions and these increased in number while in BsAs and after we left. This chapter looks at the methodology we adopted there, some initial scenario building and stakeholder analysis.

4.1 METHODOLOGY

Ferguson's post-development critique (discussed in Chapter 2) and the needs assessment activities of our 2006 Lesotho work framed WFL's approach in BsAs. Understanding the context in which we were working evolved organically, a result of the six months we spent with the cooperatives and other stakeholders that included meetings, workshops, and video and audio interviews. The filming

Figure 4.1: Caroline Baillie outside a CEAMSE social factory.

process, especially, had a multi-faceted role to play throughout the project. It served as a record keeper, as an investigative research document, as interpretive illustration, as reference material, as a problem-mapping tool and, finally, as a vehicle for story telling and knowledge sharing.

The questions that interested us and that framed our discussions during the interviews are given below. We also note the category of participant and the numbers interviewed (indicated in brackets). Where co-ops or other groups were interviewed there were usually two three members present for each. (We worked together with Rhiannon Edwards and Erika Loritz as translators when our own Spanish was not good enough.)

- Individual (10 individual/family) and group cartoneros (9 co-ops and one social factory):

    – Where do you collect your waste?

    – Which/How many hours do you work?

    – How much money do you make?

    – How long have you been doing this?

Figure 4.2: Hotpress design.

- What did you do before you did this?
- What problems have you encountered?
- What dangers do you have?
- Who owns this waste?
- Who do you sell it to?
- Do you have any funding?
- Does the government support you in any way?
- Are you part of a collective?
- Do you have a warehouse? Equipment? Gloves supplied by the government? Injections?
- Would you like to move on to manufacturing from collecting and sorting?
- Have you done this already? What is your product and process?
- What are the barriers towards you making this move?

- What materials would you use to make a product and what product ideas do you have?
- Do you have the facility here to make your own hotpress?

- Potential hotpress manufacturers (1 metal workshop manager, 2 designers and 1 recovered factory 'leader'):

  - Do you have technical skills of welding and machining?
  - Do you have access to a source of waste metal?
  - How could this press be made?
  - Do you have the electricity/access to materials and other components?
  - What would you need to make this a viable business?

- Government workers (DGPRU (2), Greenpeace member (1), CEAMSE employee (1), INTI (2), NGOs (3), micro-credit organisation (2) University employees in Buenos Aires from 3 Universities (5 - architecture, engineering and anthropology):

  - Why do cartoneros do this?
  - Where do they collect recycling?
  - Who else collects recycling? Where from?
  - What recycling goes to the recycling centre(s)
  - Who drives the recycling trucks?
  - Who owns the landfill?
  - Who picks up the garbage?
  - Who buys and sells the garbage/recycling and who do they sell it to?
  - Who does the government pay to collect the garbage?
  - What do they do with it?
  - Which/How many hours do cartoneros work?
  - What problems might they encounter?
  - Who owns the waste/recycling?
  - Who do the cartoneros sell it to?
  - Who gets funding? Why? Under what conditions?
  - How many cartoneros groups are organised?
  - How many cartoneros are there in BA?
  - Why/how does El Ceibo (the most widely know cartonero cooperative) officially collect the recycling? Could this be arranged in other areas?

Figure 4.3: Teenage Cartonero.

– How much would it cost to set up the WFL business? How much could they afford to borrow? Under what conditions can the co-ops borrow money?

In addition to the above, we were given a tour around the facilities at the CEAMSE landfill site, the social factories, the private sorting units, and the new Chinese run sorting and recycling centre. We were able to ask a limited number of questions about the cartoneros employed or to be employed in the new system.

4.2   STAKEHOLDER ANALYSIS

From the various stakeholders or "actors that can influence or be affected by a certain problem or action" (Chevalier and Buckles, 2008, p. 165), we needed to work out which ones we would work with and how. There are many stakeholder identification processes, including those developed by the Social Analysis Systems (SAS2) group (Chevalier and Buckles, 2008). They define stakeholders and point out that people may be members of different stakeholder groups. Chevalier points out that stakeholders may be identified by a number of methods including:

• Identification by experts

• Identification by self selection

- Identification by other stakeholders

- Identification using records and population data

- Identification using oral or written accounts of major events

- Identification using checklist

They use what they call a rainbow diagram to identify a small group of key stakeholders and indicate those most and least affected by the intervention or action, and those with most and least influence. We applied this to a selected group of our contacts together with an Ursula Franklin triple bottom line bookkeeping template as discussed in Chapter 1. We have worked through the potential impacts of developing the BsAs WFL initiative on various stakeholder groups.

4.3 RESULTS

4.3.1 INITIAL SCENARIO BUILDING

The first stage here is to build a general understanding of the socio-economic context as we did with Lesotho. The situation in Argentina is very different to that in Lesotho, and hence we feel it is a useful comparison to make here, so that it is possible to see how the same kinds of engineering could be applied in two very different situations but with similar socially just aims.

One consequence of the 2001 Argentinean economic crisis was massive overnight unemployment. Many tens of thousands (reported numbers vary) became *cartoneros* (literally, 'cardboard pickers'), scavenging the streets of the city of Buenos Aires for recyclable materials to sell. Even though, eight years later, their numbers have diminished, there are still an estimated 6,000 to 20,000 cooperative, family, or individual cartoneros (now referred to as 'urban recoverers') who collect, separate, sort, and sell waste as their sole economic activity (Schamber and Suarez, 2007). Some families have organised themselves into more socially sustainable cooperatives. The informal recovery of materials from waste is known to be an important survival strategy for marginalised groups in developing countries (Medina, M., 2005). Medina argues that 'waste picker cooperatives can increase the income of their members, improve their working and living conditions and promote grassroots development' (Medina, M., 2005, p. 2). He goes on to say that 'community based systems take advantage of the creativity and entrepreneurial abilities of individuals who are familiar with their communities, with the surrounding environmental and the opportunities it offers to them' (Medina, M., 2005, p. 5).

For the most part, these unpaid informal workers live in outlying shantytowns but move into the city with their carts at all times of the day, collecting and recycling an estimated 90% of whatever Buenos Aires regurgitates and finally recycles (CEAMSE, 2007). Most of the recycling waste that is collected is sold directly to agents or middlemen, though the more organised cooperatives separate, sort, and sell the materials directly to industry. Income generated by the cooperatives (approximately US $160 per month) is generally higher than that which individual cartoneros gain, though this is

Figure 4.4: Cartonero pulling cart.

still 34% below the official 2007 government poverty line of 914 pesos/month (US $245). As Medina points out, "low incomes can be explained by the low prices paid by middlemen" (Medina, M., 2005, p. 9). In rare instances, cooperatives have the machinery to reprocess the waste, fetching higher prices than simply sorted waste. The Dirección General de Políticas de Reciclado Urbano (DGPRU), grew out of the PRU (Programa de Recuperados Urbanos) when the work of the cartoneros was no longer considered a contravention of the law but was now called 'informal work.' One of its mandates was to provide credentials, gloves, tunics, vaccinations and, through registration, to legitimize the cartoneros. The office also promoted the formation of cooperatives as safer, more supportive environments for the families to work in. Children under 15 were banned from collecting waste.

Garbage collection in Buenos Aires is almost entirely privatized. The city government contracts five different trucking companies, each of which is assigned to specific city districts. Each company is responsible for cleaning the streets of waste and hauling that waste to transit points or directly to the CEAMSE (Coordinación Ecológica Area Metropolitana Sociedad del Estado) landfill. There is a single government-owned company that hauls waste from a transit point in Buenos Aires' poorest district in the southwest of the city. CEAMSE has been in business for 30 years and is the biggest player in the BsAs garbage business. It is a municipal and regional government amalgam with private affiliations, which manages the landfills that receive the waste of the 13,000,000 people of the greater metropolitan area of Buenos Aires. CEAMSE has only one landfill that is currently operational, Norte III, a little over an hour's drive from the city centre (others have either collapsed or are being turned into 'eco-parks'). CEAMSE is notably difficult to pin down, everyone having his or her opinion about the role and control of the organisation. Some say they are doing good work for the community and facilitating recycling. Others say that they are 'quasi-mafia' and will prevent any recycling scheme at all costs because it would undermine their profit margins, which are based on tonnage of garbage brought to the landfill. Buenos Aires does not yet have a visible and/or official systematic recycling program, though it's citizenry generate about 4500 tons of garbage everyday, an estimated 11% of which is disposed of by the cartoneros. The municipal government has partnered with the private trucking companies to build six sorting centres or 'green points' scattered around the city, two of which are currently up and running. Each centre will be managed by a different cooperative and receive waste hauled to it by one or more of the trucking companies. They can also receive recyclables brought to them by cartoneros in trucks (only). The sorted waste is sold somewhere up the recycling food chain and all of the proceeds benefit the cooperative. But six centres will probably not be able to support more than 300-400 cartoneros by giving them the opportunity to turn their informal work into formal work.

The question of who owns the waste is a constant theme in Buenos Aires. Residents are aware that they do not own it once they put it in the street. The private hauling companies, which own the trucks that collect the waste, believe they own it but they have little environmental incentives since they are paid to deliver all waste to the landfill. (We did witness some truckers breaking open rubbish bags to retrieve recyclables that they would sell on the side competing with the cartoneros, but there seems to be some sort of unspoken agreement as to who controls the recyclable assets of which territory.) The cartoneros who pick their way through the bags left on pavements late at night know they don't own the waste, but they are legally allowed to take the recyclables as long as they have registered with the DGPRU.

The Zero Garbage law (Basura Cero) went into effect in late 2005 with the intention of reducing garbage going into landfills or incinerators. The law stipulates that the amount of garbage in landfills is to be reduced 50 percent by 2012 and 75 percent by 2017 using 2004 levels as benchmarks. Though, on the surface, this appears to be a desirable outcome, the consequence for most cartoneros whose livelihood depends upon collecting recyclables is potentially problematic. It will be impossible for the government to professionalize the work of the thousands of informal workers and absorb them into a 'legitimate' workforce in order to fulfill Zero Garbage. This is causing strife amongst the different cartonero groups and with Greenpeace, which was instrumental in drafting the legislation and has always been a strong and dependable ally of the cartoneros. As pointed out by Medina, "scavengers respond to market forces not to environmental considerations. If there is a demand for a particular material and the price is right, they will collect it" (Medina, M., 1997, p. 25).

It is worth noting at this point that the complexities of garbage disposal, the huge amounts of money involved and the potential for organised crime is not restricted to Buenos Aires. In Giants of Garbage, Crooks et al. (1993) explores the issues of what he calls 'Big Garbage' as the "Republican growth industry of the 1990s" (Crooks et al., 1993, p. x). But as Crooks explains, as early as the 1920s the garbage industry "haulers (were) aggressively vying for each other's customers block by block and the typical contractor (was) going about his business with a horse, a wagon and a gun" (Crooks et al., 1993, p. 5). This is not the place to go into any detail about the garbage industry in 2009, either in BsAs or in the US, but it is interesting to note that some of the same key players are involved, and the growth of the few mega corporations has been spectacular. During the 1980s and under the military dictatorship in Buenos Aires, waste collection was privatised for the first time with a public/private partnership with WMI (Waste Management International), which is one of a handful of dominant players in international waste treatment and disposal. In 1990, WMI was awarded a $250 million contract for five years to 'clean the streets' of BsAs (http://www.encyclopedia.com/doc/1G1-8644848.html).

In an attempt to be responsive to the dynamics of the local context, we identified existing networks, motivations, communities, and practices as well as the tensions between these and our own ideas and needs. Our approach began with the requirement that we have no prior affiliations or associations. Consequently, we were free to explore the social and political environments from many different angles. If possible, outside funding must be cognizant of this need and allow for unfettered maneuverability. Many potential partners can be very suspicious of certain associations, especially large private corporations, NGO's or government-run development agencies, and global organisations such as The World Bank, which have particular religious, economic or political agendas. If we came to Buenos Aires with anything pre-planned, it was that we would remain as steadfastly unaligned as possible, which proved to be a prescient decision of ours given the country's long history of divisiveness. We asked questions, we spent hours on local buses, we reflected, we asked more questions. We reached out in a rhizome-like way to anyone and everyone who could help us understand the complex maze of garbage and the role of the cooperatives within that maze. We wanted to find out if the technology was useful, what products they thought could be made and sold in local markets, if the hotpress machine could be made by a local 'recovered factory' with local resources, if the money they could make would be worth the effort, and if they could borrow the money to make the press from the local microcredit organisation we discovered, The Working World (La Base). We were always questioning whether or not this was a good idea that would support the cooperatives in their struggles to stay autonomous and survive economically, or whether we should just retreat and go back to North America. This stage is an important preparatory one, which needs to be done before students get involved in the field. It can and should, however, be discussed with students as an important learning opportunity. After the initial stage, we had identified potentially interested cartonero cooperatives, university and other research partners, a microcredit organisation interested in developing a 'hotpress loan' scheme, and several roles that students could play: materials and machine development, and product development and marketing. A local industrial designer built the hotpress in Buenos Aires, which we housed at the University of Buenos Aires in Carlos Levinton's Centre for Experimental Practice in the Faculty of Architecture, Design and Urbanism. Levinton's group was using it for experimentation and demonstration to cooperative groups. We also found a local factory, 19th Deciembre (working as a cooperative or 'recovered' factory having been expropriated from its owners) that showed interested in building future models should there be a need to do so.

4.3.2   STAKEHOLDER ANALYSIS

We have applied the rainbow diagram to a sub section of our key stakeholders in the diagram below (Figure 4.5).

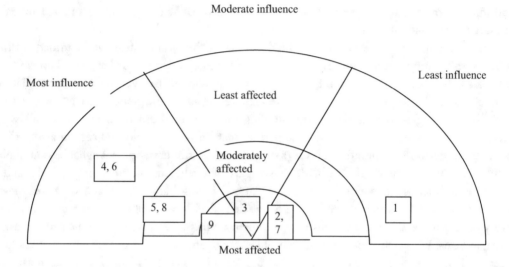

Figure 4.5: Rainbow diagram WFL Buenos Aires. 1. Street Cartoneros, 2. Social Factory, 3. Green Point cooperatives, 4. Greenpeace activists, 5. CEAMSE Landfill operators, 6. DGPRU, 7. Cartonero Cooperatives, 8. Managers/owners of Chinese sorting units/recycling factories, 9. University of Buenos Aires.

It is possible to see at a glance who are most affected or influenced by the intervention. One group of interest are the Greenpeace activists. They have a strong influence and are the ones who helped bring in the Zero Garbage Law, but are only moderately affected by WFL. It is not in their interest to know how the material is recycled or sold, so long as it does not enter a landfill. Additionally, the DGPRU are not much affected by what we do, although they may be in favour of it. CEAMSE and the commercial sorting units certainly have influence and may be affected by what we do, if successful. It would be worth watching this space if WFL begins to influence their profit margins. The University of Buenos Aires is strongly affected in a positive way as well as wielding a lot of influence. The Green point sorting units have more influence than their social factory and co-op counter parts, although this may be an illusion (see later discussion regarding Bajo Flores). They are potentially equally affected by what we do – either positively or negatively, they have the most to gain or lose. These in turn have more influence than the street cartoneros who are not affected at all by what we do.

The results of the Franklin template analysis are given in Table 4.1 below. The social and economic benefits to specific groups of people are easier to ascertain than the environmental benefits and costs as this is what Franklin calls an 'indivisible benefit' – e.g., all groups of people benefit if pollution and global warming is reduced. Instead of looking at the impact in this way, we use the idea of the groups' perceived interests in reducing environmental impact as a positive or negative benefit in the box below. Hence, the first column is only considered positive if it is in the interest of the group to reduce environmental impact. It is difficult to imagine a negative in this line. Even if a group did not want recycling to happen as it damaged their profits, that would be a negative in the economic box, not the environmental box, which would be left blank. In the table, a plus sign means positive impact, a minus sign means negative impact (strong if more than one) and a blank means no impact in this category.

Table 4.1: Franklin plot for WFL Buenos Aires.

Stakeholder	Environmental	Social	Economic
Street cartoneros			
Social factory		+	+
Green point co-ops		+	+
Greenpeace activists	+		
CEAMSE			-
DGPRU	+	+	
Co-ops		+	+
Recycling factories			-
UBA	+	+	+

Although only a rudimentary guide, this indicates that those standing to gain most from the intervention are firstly UBA and secondly the social factories, green points and co-ops. Hence, WFL Buenos Aires was launched with the notion of working with some of these groups and further detailed exploration was undertaken to explore the feasibility and impact.

These tools are useful to help analyse whether important stakeholders have been left out of discussions and also to identify those who are strongly affected but have little power to do anything about it. We will not present the outcomes of all of our data analysis in this book but will focus on certain subsections in order to demonstrate key points and in Chapters 5 and 6, we show how we worked through some critical considerations which may seem very far removed from waste recycling but which were important to understand if we were to consider working in Buenos Aires and attempting to 'socialise' our knowledge there. In Chapter 5, we focus on one stakeholder group, the local government in Buenos Aires, and its relationship with the cartonero cooperatives. We had decided by this point that if we were to go ahead, it would be with the cooperatives who were more socially and economically stable and who had in some cases moved from collecting and sorting to processing and manufacturing. These groups will be further explored in Chapter 6.

EXERCISE

Stakeholder analysis

Create a more complete version of the rainbow diagram with all actors you can find within this text.

CHAPTER 5

Stakeholder Focus: The Local Government

In this chapter, we look in some detail at the role of the local government in relation to the cartoneros. Were they shirking their responsibility for waste recycling by allowing the cartoneros to do this work, or would they prefer it if cartoneros left the city and did not challenge their desired status of a developed nation? The changing socio-economic climate in Argentina from the early 1990s has dramatically affected the nature of the country's informal economy and, consequently, the government policy towards it. With estimated figures of between 6 and 20,000 informal urban recoverers operating in the city of Buenos Aires alone (see below), the municipal government of Buenos Aires has been forced to address the situation with the creation of suitable policies. However, little is understood of the impacts which these policies have on the realities of the individual cartoneros or urban recoverers they target, rendering assessment of how suitable they really are and further creation of appropriate policies impossible. In order to understand our context in Buenos Aires better, in this chapter, we consider the question: 'What actions have the Municipal government of Buenos Aires taken with regards to the city's urban recoverers and what were the perceived impacts of these actions?' But first, we must step back and look at the idea of government policy in relation to what is referred to as the 'informal economy.'

Since Keith Hart conducted research into informal income generation in Ghana in the 1970s (Chen, M., 2007), the study of so-called 'informal economies' has been prevalent in many areas of academic research and government policy-making. Modernisation theorists viewed this form of unregulated work as an aspect of 'third world' countries that would ultimately disappear as development occurred (Chen, M., 2007). When this did not happen, debates began over: a) Whether informal work was in fact something that could promote development rather than acting as a sign of its absence, and b) How this should affect policy-making (Chen, M., 2007).

It is evident from attempts to define the term informal economy, that it is a heterogeneous phenomenon encompassing a huge number of income-generating activities. The very lack of government regulation that causes some to define it as informal (Marcouillier et al., 1997) leaves it open to many different methods of organisation. One such example that has occurred internationally is the cooperative movement, about which there is a large amount of conceptual literature concerning the sociological and anthropological perspective on cooperativism and solidarity economies. These typically focus on the nature of societies in which cooperatives develop and how they tend to operate, examples being works by Jossa, B. (2005), Defourny and Develtere (1999) and Smith, M.

(1988). Jossa, B. (2005) examines co-operatives' potential as a force for encouraging equality from a Marxist perspective, giving an abstract and historical account of how they have developed in society. Defourny and Develtere (1999) and Smith both give a conceptual analysis of the ideological forces that drive cooperativism, such as religious beliefs and economic necessity. They also assess the general conditions that are needed for cooperatives to develop, with Smith, M. (1988) specifically considering this within the context of behavioural structures in Latin American society.

Mendieta, E. (2006) focuses on meta-urban issues in Latin American cities, with reference to the development of an informal recycling sector in Argentina after the 2001 crisis as a way of generating income and as a statement against neoliberal economic environments. Chronopoulos, T. (2006) provides a historical timeline of the emergence of informal recycling workers, or cartoneros in Buenos Aires, combined with primary ethnographic case studies about the type of work cartoneros do and the socio-economic context in which they operate. This includes reference to a number of specific government actions, particularly after the economic crisis, with analysis of impacts primarily focusing on income generation. As with Mendienta, Chronopolous provides a top-down, structural perspective of informal recycling as a livelihood, concluding that it is a positive 'solution' to the problem caused by neoliberal economic policies. However, neither considers the impacts of government decisions at an individual level or the diversity of needs within a country's informal economy. In this part of our background stakeholder analysis, we began to explore these relationships in some detail as we could see how critical they would be in respect of the sustainability of WFL BsAs and the choice about who we would partner with.

5.1 METHODOLOGY

Due to the connections this analysis makes between existing government policy and the reality of its impacts and in order to develop an understanding of the actors involved, during selected interviews described in Chapter 4, extra questions were asked related to this particular focussed study. Furthermore, data was gathered from a range of sources at different levels. Sources used include government publications, academic literature and news items. The initial questions asked are set out in Table 5.1.

An overview of those who were asked these questions are given in Table 5.2. For reasons of confidentiality, they have been identified with codes throughout. However, for the purpose of clarity of writing, the gender of each participant has not been concealed as it is not deemed to breach confidentiality. Some groups were not interviewed but observed only (Table 5.3).

Table 5.1: Interviews with Urban Recovers.

	Questions	Anticipated Information
1	Who are they?	Name, position, other personal details e.g., age and sex
2	What is the nature of their work?	Activities involved, working hours, location of work, income
3	Can they identify government actions regarding their work?	Prior awareness of some government actions from literature that may be mentioned
4	What do they perceive the impacts to be?	This will depend entirely on individual circumstance. However, supportive government action is expected to be viewed positively, and restrictive action negatively.
5	How has their work changed?	As above
6	What do they hope/predict for the future of their work?	Greatly dependent on answers to 3, 4 and 5. Opinions of incoming government.

5.2   DATA ANALYSIS AND RESULTS OF THE STUDY

Because this research combines an understanding of actors working at different levels, the approach to data analysis was inspired by Dey, I. (1993), allowing for a holistic understanding of the issue without neglecting detail. In order to achieve this, it was necessary to organise the large amounts of data collected into meaningful areas which could be analysed separately and in relation to one another.

Interviews were transcribed from dictaphone recordings, then the data was divided into themes or categories from which information could be studied individually or in relation to others. The pieces of data put into each theme were defined in numerous ways; for example, some direct answers to the initial questions, some touched on key points in the literature, some issues that were raised frequently by participants and some that participants appeared to feel particularly strongly about. There was no set way of defining what was important and what was not; the process of analysis was conducted by sifting through the data and pulling out different aspects. Once in categories, the data were analysed in relation to other information in the same category, different categories and in relation to government policies and relevant literature as shown in Table 5.4. Details of the data analysis pertaining to each question are given below.

Table 5.2: Interview Participants.		
Participant Identification	**Relevance to Research Question**	**Location**
Participant A	Part of worker's co-operative, the Unión Solidaria de los Trabajadores (UST) 'Workers' Solidarity Union.' Villa Dominico.	UST Villa Dominico site.
Participant B (Translated)	Founder and leader of plastics recycling cooperative.	Co-operative site
Participant C (Translated)	Independent urban recoverer. Collects all recyclable items in neighbourhood to sell and works with co-operatives.	Participant's home and workplace.
Participant D (Translated)	Ex-cartonero and members Bajo Flores recycling co-operative.	Bajo Flores plant.
Participant Name	Role within Context of Research	Location
Participants 1, 2, and 3 (2 and 3 translated)	Members of previous (current at time of interview) municipal government's department of Environment and Public Space, Development section.	government Offices
Participant 4	Researcher at Buenos Aires University responsible for conducting studies into the situation of urban recoverers in the city.	University of Buenos Aires offices
Participant 5	Member of Working World; an organisation that provides credit for co-operatives in Buenos Aires.	Residence, Defensa, San Telmo, Buenos Aires

5.2.1   WHO ARE THE URBAN RECOVERERS OF BUENOS AIRES?

Definition of the Term 'Urban Recoverer?'

Buenos Aires is home to many people informally collecting, processing and selling recyclable materials, but despite often sharing a common term by which they are identified, these people perform a diverse range of activities and can be divided into many social and economic categories. In literature regarding informal waste collection, numerous terms are used, including 'scavenger,' 'waste picker' (Medina, M., 2000), and 'ciruja' (Berger and Blugerman, 2006). Chronopoulos, T. (2006) uses the term 'cartonero' to describe individuals who collect recyclables from the streets in order to process or sell them. This is the name most frequently used by the media and the government, derived from the material that in the past was most commonly collected, cardboard, or cartón. De-

Table 5.3: Observed Groups.

Observed Areas	Relevance to Research Question	Location
Observed A	Small plastics recycling co-operative	Villa Angelica
Observed B	Highly mechanised plastics recycling co-operative	Esteplasm

Table 5.4: Categories emerging from the interviews.

Theme of Chapter	Issues Raised
1. Who are the urban recoverers of Buenos Aires?	Definitions of the term 'urban recoverer' Demographic composition of the informal recycling industry The role of cooperatives
2. Who 'owns' waste?	The formal waste collection system The role of the private sector The role of the informal sector
3. Contextual factors affecting government action	Conceptual economic context of government policies Historical context of government policies
4. Formal government actions and the reality of their impacts	Recycling policies: formal government action and perceptions of them Policies of organisation and perceptions of them Health and social policies and perceptions of them
5. What do urban recoverers want from the government?	Urban recoverers' perceptions of the concept of regulation and government action

spite the frequent use of the term cartonero, the official name given by the municipal government and also used by most participants during the fieldwork is 'urban recoverer.' This term accounts for both independent and co-operative workers who make their livelihood from the informal collection, processing or selling of recyclables in the city. Participant C, an urban recoverer in Buenos Aires, explained that to her the term cartonero referred specifically to someone collecting only cardboard and is not a term recognised by law. The formally recognised term of 'urban recoverer' gave her a sense of legal status. The name given to this activity is the first of a number of examples of inconsistency between the actors in this system. The municipal government's website also uses the term 'urban recoverer,' but during a meeting commissioned by Chief of government Mauricio Macri to discuss informal recycling, all actors present including government and NGO representatives used the word cartonero. We will continue to use the word cartoneros throughout this text but wish to acknowledge the sensitivity surrounding this word.

Demographic Composition of Informal Recycling Industry

The measurement of the scale at which informal recycling operates in Buenos Aires suffers from discrepancy, because of the following:

a) There is no easy way of monitoring informal, unregulated work,

b) It is not necessarily 'regular' work for participants,

c) The figures are likely to have altered dramatically over recent years due to a changing political, social and economic context.

Conflicting knowledge about the nature of urban recovery was a consistent theme during the fieldwork. Examples of inconsistency between figures gained from literature and the fieldwork are shown in Table 5.5.

Table 5.5: Estimates of numbers of urban recoverers in BsAs.

Source	Number of cartoneros
Fieldwork: government meeting commissioned by Macri	20-30,000
Fieldwork: Participant 1	9,000 registered. 5,000 regularly work.
Fieldwork: Participant 4	18,000
Anthropologist Francisco Suarez (CNN, 2003)	10,000

Participant 1 explained that the government's difficulties in including cartoneros in their waste collection and disposal programmes were often based on a lack of understanding about the nature of their work and their needs. Although 9,000 are currently on the government's official register, this registration only lasts for a year, and they suspect many do not re-register if they do not feel the process has been beneficial. She identified the informal nature of this work as a key reason for their inability to quantify it, as sometimes cartoneros would find a short term contract in the formal sector, and then return when it finished. Others would have formal employment and only collect or process recyclables to supplement their regular income.

The Role of Cooperatives

The informal recycling industry boomed at the same time as another grassroots movement in Argentina: the development of co-operatives as a means of replacing the vacuum left by neo-liberalist government policies (Neamtan, N., 2002). According to Defourny and Develtere (1999), cooperative movements occur through necessity, and their ideological grounding helps them operate effectively. Participant 3 explained that this necessity was created by Menemist neoliberal policies in the 1990s and diminished institutional support after the economic crisis. He used 'recovered industries,' such as the Hotel Bauen and textile factories in Buenos Aires as examples of this situation, where workers who were left unemployed after their place of work closed down and the owners went bankrupt, chose to go back to work without a manager and run the business themselves. The participant explained that the success of co-operatives and expansion of neighbourhood assemblies resulted in strong social networks that saw the potential that working together held. Urban recoverers started working together in order to provide each other with some financial stability, better economies of scale and increased bargaining power (Berger and Blugerman, 2006). UST is one example of a successful co-operative, comprising 140 workers who had previously been contracted by CEAMSE to collect and dispose of urban waste from the city of Buenos Aires. When the contract was not renewed in 2002, the workers formed a co-operative, but despite their 24 years of experience in the job, CEAMSE elected to employ less costly labour from abroad. Many months of legal battling ensued and was only resolved when CEAMSE was pressured by the workers, with the assistance of the Minister of Labour, to settle the issue, after which, they contracted UST to perform post-closing maintenance on one of their landfill sites.

'We workers can take our destiny in our own hands' [sic] (UST, 2007, p. 4) leaflet.

5.2.2   WHO OWNS THE WASTE?

The question of who owns the city's waste has been raised many times both in literature and during the fieldwork. It received national press attention when Mauricio Macri, the current Chief of government in the city, stated during his 2002 election campaign that cartoneros were stealing waste, a sentiment reflected in the press (Chronopoulos, T., 2006).

This issue is central to the types of action the municipal government has taken regarding cartoneros, as their role in the waste collection and recycling industries has changed over the past decade. Media coverage and public opinion have played a major part in these changes.

The Formal Waste Collection System

When asked who she thought 'owned' waste during interviews, Participant 1 explained that there are two descriptions one can give of waste collection and recycling in Buenos Aires, the formal and the informal. The first actor to consider in the formal system is the city government. In 1994, the autonomous capital city of Argentina (Ciudad Autónoma de Buenos Aires, or CABA) was granted equivalent institutional rights to the other Argentinean provinces, with its own constitution and Chief of government, or 'Mayor' (Pirez, A., 2002). In February 2007, Mauricio Macri was elected as Chief of government with his neo-liberal 'Commitment to Change' party. Within his government, the departments that have the most influence over waste management and recycling are found within the General Directorate for Urban Hygiene, headed by Gustavo Grasso, and the subdivisions within, as shown in Figure 5.1.

The Role of the Private Sector

Although the Urban Hygiene Department in Buenos Aires is responsible for the organisation of waste disposal, in reality, they only directly serve one 'zone' of the city, and delegate service provision to five private companies in the other zones, shown in Figure 5.2.

The contracted companies shown in Figure 5.2 are part of the next stage of the formal system, the private sector. The municipal government's official publications do not clarify their relationship with the private sector and Participants 1, 2, and 3 explained that they found it ambiguous. However, during the fieldwork, it became clear that private companies played a major role in government policy and its impact.

When asked who owned waste, Participant 4 explained that, initially, it is whoever produces it. As soon as waste is taken at street level the municipal government owns it, as they are responsible for collection through their contracted companies. The private trucking companies move it to what is called a 'transfer station,' at which stage, it belongs to the Coordinación Ecológica Área Metropolitana Sociedad del Estado (CEAMSE); the Metropolitan Area Ecological Coordination, National Corporation, who take it to their landfills. Participant 4 explained that CEAMSE is technically a government agency, created by the military dictatorship in 1978. Now they are a combination of members of the municipal government and private companies, all funded by the government to dispose of waste in their landfills.

The complex nature of 'who owns waste' and more specifically 'who owns recycling' is part of the difficulty the government faces in their approaches to the informal sector.

The Role of the Informal Sector

Participant 3 stated that government estimates show that cartoneros recoup 11% of the city's waste and 97% of recyclables. In comparison, official schemes recoup only 1 or 2%. Usually recycling collected informally is sold by individuals or co-operatives, either before or after processing, to 'galpon depositos,' which are small private or co-operative run storehouses. Here it is separated, washed and processed, after which, it is resold by type of material to bigger industries.

Despite their successful operation, according to Macri's statement, at no stage do these cartoneros have the 'right' to take and sell waste. According to Participant 4, Macri's interest in the waste collection system is made all the more ambiguous by the fact he owns one of the private collection companies who are now also setting up profitable recycling programmes.

5.2.3   CONTEXTUAL FACTORS AFFECTING GOVERNMENT ACTION

Historical Context

The 'economic crisis' of 2001/2 has undoubtedly had an enormous impact on cartoneros, the role of co-operatives, the informal economy and government policy. The recent history of the financial

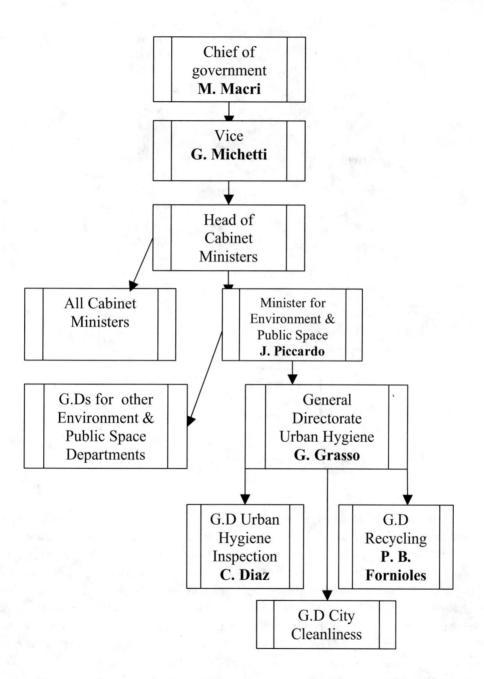

Figure 5.1: Structure of government departments responsible for waste management (GCBA, 2008).

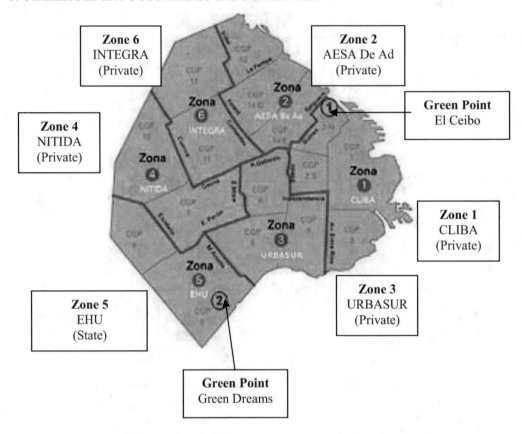

Figure 5.2: Buenos Aires waste collection zones and Green Points (GCBA, 2008).

collapse can be traced to the dictator Jorge Rafael Videla, who took out large loans from international institutions in the 1970s (Cibils, A., 2006). The national government 'pegged' the peso to the dollar in 1991, resulting in a high rate of exchange, which damaged Argentina's export industry as goods became more expensive than those from other Latin American countries with weaker currencies (De La Torre et al., 2002). Participant 4 explained that her studies at the University of Buenos Aires showed the cost of paper rose from $90 (pesos) to $270 per ton, thus resulting in further unemployment and making the recycling industry far more lucrative. The neoliberal economical reforms that were introduced by President Menem during the 1990s to strengthen the economy included massive privatisation of national industries and deregulation of economic activities, allowing freedom of interaction with international markets (Chronopoulos, T., 2006).

By the late 1990s, it became increasingly clear to the Argentinean government and the international community that Argentina would not be able to service its debts. A Structural Adjustment

Figure 5.3: Sorting recybclables at CEAMSE.

Programmes (SAP) known as Blindaje was designed by international lending institutions to cut public expenditure, which included major cutbacks of state provisions and further neoliberal reforms. Chen et al. (2001) state that recognition of informal activities as an enduring feature of modern capitalism is vital if governments are going to take appropriate actions towards them. In Buenos Aires, informal recycling ceased to be considered a marginal activity during the 1990s as the numbers of cartoneros increased, despite their illegal status since the 1980s until recently (Berger and Blugerman, 2006). This continued after the 2001 economic crisis as a result of the media attention given to the plight of cartoneros. A theme that arose constantly during the fieldwork was the changing image of informal recycling in the public eye and how it affected government policy. Participant B identified the difference he had been aware of between the image of co-operatives like his before the crisis and after. The main difference, he said, was that they gained public sympathy as people realised how easily they could find themselves forced into the same unstable situation.

'In the last few years, at a national level, has occurred a national recognition …not only from the government but also by neighbours who are now separating waste.' (Participant B).

'…this government pays attention to public opinion and the media, and the media is made, as in any country, by the middle classes, not the poor.' (Participant 2).

The public and, consequently, the government were alerted to the negative aspects of urban recovery, including its vulnerable economic position and lack of social support structures. As a result, laws and policies were created in order to address these issues. Participant 1 confirmed this, but she went on to say she believed the times of sympathy towards cartoneros passed as soon as people found themselves in more stable financial positions and, consequently, lost interest in their plight.

'When that fear went out, when middle classes in Buenos Aires saw they had their own place, that they felt they were not going to be on the streets at any rate, all that understanding of cartoñeros slowly went down.' (Participant 1).

As public and media interest diminished, so did pressure on the government to take action to support the cartoneros. Mauricio Macri's successful 2007 election campaign may be indicative of the change in public opinion, given his attitude towards informal workers in the past.

Conceptual Economic Context of Government Policies

Due to the complex nature of waste collection and recycling in Buenos Aires, along with economic and social issues raised by the existence of an 'informal' workforce, the municipal government is attempting to implement greater regulation. At this point, it is necessary to bring together the economic concepts of how a government should regulate informal activities with the reality of the Buenos Aires government's methods of intervention, in order to assess their impacts (Table 5.6).

Table 5.6: Conceptual economic views of government action in the informal sector (Portes and Schauffler, 1993).

Conceptual Approach	Key Elements Relating to the Informal Economy
Dualist	Identification of the informal and formal economies as two distinct entities with limited interaction Less concerned with government regulation of the informal sector, and more with available business and social support programmes
Legalist	Advocates economic deregulation based on the damage state-regulation can do to informal workers' abilities to move into the inaccessible and self-preserving formal arena
Structuralist	Recognises intrinsic and potentially damaging links between formal and informal economies Focuses on the responsibility of the government to intervene and protect vulnerable informal workers at risk of exploitation or marginalisation by private corporations

In the case of informal urban recovery in Buenos Aires, the de-regulative strategies the municipal government had used in the past could certainly be argued to have increased the size of the informal sector. However, more recent changes in Argentina's political and economic environment appear to have resulted in a greater degree of regulation and intervention with the aim of aiding informal workers.

5.2.4   FORMAL GOVERNMENT ACTIONS AND THE REALITY OF THEIR IMPACTS

Recycling Policies

In recent years, the municipal government has been taking action to reduce waste production by promoting recycling. The Ministry of Environment and Public Space's Urban Hygiene Department has prepared numerous integrative strategies to deal with high levels of waste generation in the city and accommodate for all actors (Table 5.7). Major changes to their waste collection and disposal system started in February 2005 when the previous administration established a strategy called Gestión Integral de Residuos Sólidos Urbanos (GIRSU) or Integrated Management of Urban Solid Waste, which the current government has continued (GCBA, 2008).

These policies are attempting to make Buenos Aires an environmentally friendly city, but they also have the potential to cut out the role of cartoneros by making private collection companies responsible for gathering recyclables. The issue of whether or not the government is trying to 'formalise' the currently informal stages of waste collection and how cartoneros will fit in was an enormous source of contention throughout the fieldwork. Paying collection companies to keep their sector 'clean' has eliminated the competition that used to exist between them and cartoneros over recyclables. However, Participant 3 stated that problems have arisen due to poor monitoring of the system and lack of clarity over what is meant by a 'clean area.' Participant 1 said that government statistics show that informal collection accounts for 97% of the recovery of recyclables, and official collection such as Recolección Diferenciada just 2%. She also stated that, '…only 30% in recycling containers is recycling; the other 70% is common garbage.'

It seems that the current informal system is effective in reducing waste being sent to landfill sites. Legalist opinion would argue that this is grounds for reduced state intervention into the informal sector, as informal industries will adapt to local needs in order to survive, and are often extremely efficient as a result.

Policies of Organisation and Perception of them

Despite cartoneros' major contribution to the recycling system, their informal nature has often been viewed as a problem rather than solution to waste collection and disposal, according to Participant 3. At the meeting organised by Macri, the 'problem of cartoñeros' was constantly referred to, along with the 'solution' that could be provided by attempting to formalise parts of their work. This was also the main priority of Participant 4 and cartonero Participant C, although interpreted in different ways. Participant 4 advocated putting them into the formal sector in order to increase their job security, by employing them in Green Point transfer zones and allowing private companies to do the collection. 'The thing I want is for cartoneros to have more sanitary, legal work…with security. Put them in the formal sector.' (Participant 4).

Name	Key Operational Elements	Overall Aims
Recolección Diferenciada, (Waste Separation)	Provision of separate collections for recyclables by private collection companies	To decrease waste generated in the household To only send waste with no further value to landfill sites To encourages residents to separate waste at home
Sector Plan	Provides government field operators to target each of the six collection zones' specific needs. As a result some areas are given communal recycling bins and some have had stricter regulations imposed on them regarding how much of their waste can be disposed of in landfill sites. Rather than receiving payment per ton of waste collected, companies are paid to keep their sector clean.	As above Change in payment structure allows cartoneros and private companies to work together, rather than against each other
Zero Waste law	Legislation enacted May 2007 as a result of Greenpeace lobbies	Prohibiting the disposal of recyclable products in landfill sites by 2020

Table 5.7: GIRSU Recycling strategies (GCBA, 2008).

Participant C promoted a greater degree of formalisation by having all cartoneros working in co-operatives and setting up official door-to-door collection within their neighbourhood to ensure they are not cut out of the system, stating

'It is fundamental that separation is done at home, and cartoneros collect it directly.'

Despite recent neoliberal economic history, the municipal government has introduced a number of policies to lessen the degree of informality in cartoneros' work, or at least mitigate its negative aspects. The 992 Law was a strategy frequently mentioned during the fieldwork. The key elements of this law are shown in Table 5.8.

Participant B claimed that as a result of legalisation policies, cartoneros had,

'…greater recognition in the last few years. We are no longer thought of as thieves.'

Table 5.8: Government organisational strategies (GCBA, 2008).

Operational Factors	Overall Aims
Legalisation from the 992 Law. Registration of recoverers through, 1) Registros de Recuperadores Urbanos (RRU), Register of Urban Recoverers and 2) Registro Permanente de Cooperativas y Pequeñas y Medianas Empresas (REPyME), Permanent Register of Cooperatives and Small and Medium Enterprises	To increase awareness and subsequently status of urban recovers through public education To legalise the activities of cartoneros over the age of 18 and therefore reduce child labour forces To monitor the nature of urban recovery activities To provide social support for informal workers, including healthcare and training To give greater 'ownership' of waste to cartoneros (Participant 1)
The five private and one public collection companies taking separated waste from households and communal bins and delivering it to one of six Green Points. In the Green Points cooperatives of cartoneros will separate and process recyclables and sell them on to industry.	Encourage greater levels formalisation through the increased formation of cooperatives. Provide safer working environments. Ensure cooperatives' access to materials, stabilising their activities

Chen et al. (2001) claim that legal registration and ID cards give informal workers a right to work, along with a sense of legal identity in the economy. This is one way of improving government understanding of work that is done informally, but it is not necessarily effective in many cases due to the inconvenience caused to informal workers of completing registration. Chen et al. note that spending time (up to four hours according to GCBA (2008)) queuing to register when they could be working or the process of filling in complex forms are not necessarily suitable to all members of heterogeneous informal systems.

Green Points as a way of organising and giving a greater degree of formalisation to urban recovery were the subject of major criticism. Participant 2 stated that the weakness of official recycling programmes made the Green Points unrealistic. 'The cartoneros gather [at Green Points]. But we find a problem – the materials gathered by Recolección Diferenciada is very small. In the near future, will the government start collecting 100% of recyclables? We doubt it.'

Along with the question of how Green Points would receive materials was the criticism that even if they were to work at full capacity, they would not be able to accommodate every cartonero. Participant C expressed her concerns that, 'the Centro Verdes won't be able to absorb the number of people needing work…and the rest wouldn't even have the garbage to pick up [from the streets].'

Participant 4, an advocate of attempting to organise and stabilise urban recovery through government policy, was aware that in Green Points 'It would be possible to employ not all, but some yes' but felt that this was better than the current system of ignoring the situation and allowing all cartoneros to work under difficult conditions. A similar concern was raised over the government's expectation that all cartoneros could be encouraged in this way to operate in cooperatives. Participant 1 claimed that this was 'impossible and wouldn't work. City government policies that think this is possible are lying to themselves. It would be great…but is not possible.'

Chen, M. (2007) identifies the arguments within literature that this kind 'formalisation' of informal work, whether desirable or not, is simply not feasible based on a) the bureaucratic costs it would pose to governments of funding, paperwork, etc, and b) the fact that working age populations in developing countries are typically large, and therefore, formal sectors could not realistically provide enough employment. Participant 1 claimed the result of Green Points would be that 'There would be nothing left [for the cartoneros]. It is not so easy to say 5,000 people can do the same work as now, but in a formal way. It's just not possible! It's difficult to accept, but the way they do it now is the only efficient way it can be done.' The practicalities of setting up and running Green Points were another area of criticism. Participant 4 claimed that only one had ever been operating, in contrast to government reports of two, as shown in Figure 5.2. She went on to say that the one operating Green Point was closed due to a fire '…and now [the waste] goes to this warehouse, waiting.' It appears that, as with other policies, this system is already suffering from weak implementation and delays.

The role of private companies within the proposed system was a source of contention, with suspicions raised as to whether truck drivers from the collection companies were taking recyclables, explaining why Green Points (or the warehouses) received so little. Participant C believed that 'what should go to the Green Points is taken by the truck drivers.' This returns to the question of 'who owns waste' as although the truck drivers are responsible for the materials during transportation, they are contracted by the government to take it to the Green Points for cartoneros. Participant 1 explained that often the trucking company workers would strike informal deals with cartoneros over the 'ownership' of recyclables. The cartoneros have limited bargaining power due to their relative disorganisation and the informality of their work, so often agree to give them access to some recyclables for private use.

During the fieldwork, three questions were raised regarding the government's attempt to formalise, or at least stabilise informal recycling. These were the following:

- How will Green Points receive enough materials to support them if cartoneros are no longer able to collect from the streets?

- What will happen to the informal workers that are not accommodated for in six Green Points?

• What are the dangers posed by weak implementation and regulation of the project, and can they be overcome?

Chen, M. (2007) raises the question of how feasible 'formalisation' is, given the costs this would incur to the government and the difficulties they have experienced in monitoring informal activities. De Soto, H. (1989) would argue that this is where government policies that forcibly formalise work will ultimately exclude anyone who does not fit into the policy, which certainly appears to be the case with Green Points. At the Macri-commissioned meeting, Green Points were seen as a key future solution to the 'problem,' but with their estimated figure of 20,000 cartoneros in the city, there was no mention of how they would be accommodated for in a maximum of six Green Points, which they estimated would employ 200 workers in each. When asked for her views, Participant 1 stated that 'this question hasn't been answered by anyone. No group, person or place has the answer.'

Health and Social Policies and Perceptions of Them
Other than encouraging greater economic stability, the government is attempting to address the lack of social support cartoneros receive and the personal dangers they face. This issue has remained central to public concern since 2001 and may serve as an explanation as to why the government is implementing seemingly unrealistic formalisation schemes. Table 5.9 shows the government strategies in place to address social issues.

At the meeting commissioned by Macri, the main issues raised with regards to informal urban recovery were poor working conditions, poor sanitation and child labour. They highlighted the need for social development programmes that specifically target these problems, although they were not specific about what these should entail. Participant B identified the dangers faced by cartoneros working on and living near landfill sites. Contamination of their water supply from the landfill resulted in increasing cases of disease, and it increased risk of cancer from exposure to toxic waste. Participant 4's studies showed that people relying on the landfill as a source of recyclable materials are moving closer to it every year in order to decrease travel time, increasing exposure to contaminated water and poisonous gases, 'In 1972 there was a 3 km buffer zone, but they get closer and closer.' Her investigations have shown the landfill sites where informal collection is conducted contain traces of e-coli, hepatitis and many other diseases. There had also been sightings of people picking up discarded food to eat. Observations of plastics processing cooperatives revealed the dangers posed by a lack of enforced health regulations in the workplace. The warehouses were filled with dust from the grinding up of plastic items into chipping in order to be resold. Workers were not wearing protective masks, and there was little evidence of ventilation systems. With regards to child labour, Participant 4's studies showed that 50% of working cartoneros are under the age of 15.

Participant 1 explained that the health issues posed by working on landfill sites were brought to media attention when a cartonero died from a fall whilst collecting on one of CEAMSE's landfills. 'They [CEAMSE] used to have problems with cartoneros going onto landfill sites, including one fatality, so they were made to do something.'

Table 5.9: Government strategies to address health and social issues.

Strategy	Operational Elements
992 Law	Training programmes regarding safety at work
	Protective clothing and gloves
	Vaccinations against the most common diseases exposed to at work
	Making child labour illegal
CEAMSE Social Plants	Discourage dangerous and unsanitary collection on landfill sites by providing alternative workspaces

The 'formalising' aspects of plans to improve health and social support suffer familiar criticism. CEAMSE's social plants raised the question of what would happen to the recoverers not employed by the plant, as an estimated 1,500 people collect daily from Norte III alone (Participant 4). The visit to the Norte III CEAMSE landfill in the province of Buenos Aires gave an insight into how these plants are working, with one social plant and one private plant currently in operation. The private recycling plant was highly mechanised and required limited labour, so a social plant had been set up nearby to employ urban recoverers who used to collect on the landfill. There was a new high-tech Chinese plant under construction, which when completed would process enormous amounts of recycling by machine. When asked what would happen to the workers in the social plants or those who still go onto the landfill when the Chinese plant was complete, CEAMSE representatives failed to give an answer, claiming not to understand the question either in Spanish or English.

Despite these strategies, CEAMSE still allows urban recoverers onto the landfill every afternoon between three and five o'clock. Participant 2 claimed that in terms of social impacts the 992 Law had been one of the most appropriate and successful government schemes, as shown by the impacts on co-operatives such as the one run by Participant B, who were given protective clothing and the opportunity for ten members to attend training courses that informed them on different aspects of their work. 'We took courses, went to workshops and conferences about theory and practise, the whole process of plastics. From how it is formed, to the dangers of it, such as what to do if a machine catches fire.' (Participant B). With regards to the issue of child labour, Participant 4 claimed that although the government had created appropriate laws, not enough was being done to implement them. When asked if action was being taken to encourage under-15s back into education, she said, '[The government] take action, but it is not the action they need to do. They need to oblige children to go to school, for example, by paying them.' It seems that until families have a sense of financial stability a law is unlikely to discourage family members from working as urban recoverers, whatever the disadvantages of the job may be.

5.2.5   WHAT DO URBAN RECOVERERS WANT FROM THE GOVERNMENT?

Within the discussion about what is 'best' for informal workers, between safer and more stable formal work and the current informal system that includes everyone but has its costs, one voice is rarely heard. Urban recoverers suffer from a major lack of representation, as displayed in the meeting commissioned by Macri which was designed to bring together 'all' actors within the system to discuss appropriate action. Their final conclusion was that the situation could only improve, i.e., through introduction of formalisation projects such as Green Points) with an integrative approach from the bottom-up. However, despite the presence of members of the incoming government, journalists, high court judges and NGO representatives, there was only one person with experience of urban recovery, an ex-co-operative member. He expressed his discomfort with this situation, to which the discussion leader responded that including urban recoverers in such meetings was very difficult due to their lack of organisation. This made it hard to contact them or arrange meeting times.

Along with the problems caused by a lack of representation of urban recoverers in policy making, there is also the issue of what kind of government intervention urban recoverers want. The structuralist view that government regulation should protect informal workers (Chen, M., 2007) suggests that government intervention will ultimately, a) have a positive influence on livelihoods and b) reflect the wants of individual members of the informal sector. However, interviews showed this is not necessarily the case. Participant 5 explained that in his work, he had seen continuing distrust of the government as urban recoverers had lived through military dictatorships, ineffective economic policy and subsequent collapse, and accusations from government members of their 'stealing' of waste. Participant A expressed his pride in their co-operative's ability to function by itself, without government intervention. The nature of the co-operative movement is based on not needing higher forces to organise a livelihood for them; they are able to provide for themselves through cooperation.

Participant 5 noted that trust in the government has been further diminished by the weak implementation of their policies, for example, the unfinished Green Points and lack of plans for what will happen after they are built. They also have a track record of offering subsidies to cooperatives and failing to produce them. '…[cooperatives] spent maybe a year or two waiting for a subsidy they thought they might get in a month or two. Then only one in ten got one at all.' Participant 5 claimed that struggling co-operatives often find it difficult to resist the offer of government aid even if it resulted in the suppression of the movement's political ideology because their necessity to generate income has to prevail. He described it as 'a 'normalising technique' the government could use to effectively suppress their 'progressive energy,' i.e., after two years of waiting their ideological ambitions have been replaced by their necessity to survive.

Participant D confirmed these sentiments when asked what he thought of the government's action towards recycling co-operatives such as his. He explained that they are given a subsidies every so often, but they are only ever enough to 'cover holes' and they would prefer larger amounts of credit to make a significant improvement to their business. Participant D explained that the main problem he had with the government was its lack of continuity, with different policies and projects constantly being introduced but never completed. When asked what he wanted from the government other than the option of credit he replied, 'We are entirely independent from the government. Everything we have achieved is entirely down to ourselves, not because the government has organised them but because we work together.'

Participant B appeared to be content with the government support they had received, although they would have preferred greater support with their productive rather than social development. Participant C claimed that the government policies, although not being particularly appropriate to the needs of the urban recoverers, were not damaging either, particularly ones such as Recolección Diferenciada, which saved recoverers from the unsanitary work of rooting through bin bags. However, in order to prevent the attempted formalisation of the system from cutting out informal workers that didn't fit into Green Points, she claimed they needed to organise themselves better.

'According to the level of organisation [government initiatives] wouldn't affect them, because if you go door-to-door and they give [recycling] to you directly, it won't get put in the recycling bins.' (Participant C). This would allow them to formalise, but in such a way that made them officially responsible for more of the system. However, she warned that if this did not happen.. '…there are still lots of un-organised [urban recoverers], so yes, it will put them at a disadvantage.' (Participant C).

It appears that there is a conflicting response to government policy at ground level, with lack of faith in government policy and cooperative ideology creating an apathy towards, or even negative view of any government intervention. This is either due to the problems caused by specific policies, such as the risk of Green Points cutting informal workers out of the system, or a more ideological desire to operate under minimal state control. Despite conceptual debates over the ideology of cooperatism and how it fits in to a legalist or structuralist economic approach, in reality, opinions appear to be more a matter of practicality. Workers recognise the potentially damaging impacts of government policies at ground level or lack trust in a government that has failed to support them in the past and has consequently been replaced with the support they receive from cooperatives.

5.3   CONCLUSIONS

From a theoretical standpoint, urban recoverers have been affected by changing scales of regulation of Argentina's economy over the past decade. As expected by structuralists, with the decreasing role of the state the informal recycling industry grew dramatically in size. As expected by legalists, these neoliberal policies encouraged economic growth which provided greater economic choice and freedom. The informal sector, once freed from the legislation that had made urban recovery illegal for three decades, started operating in an efficient and cost-effective manner. It absorbed many people in need of economic support that the state could no longer provide and also transformed the city's waste management operations, reducing industries' reliance on expensive imports, thus making exports more lucrative. From an environmentalist perspective, informal urban recovery became a sustainable option within waste disposal at no cost to the government.

However, despite the efficiency of the system from an economic perspective, a combination of public concern for the social impacts and increasing recognition of informal workers' vulnerability with regards to private collection companies resulted in an increasingly regulated approach by the city government. They created a number of policies aimed at organising and regulating the informal recycling sector, as well as taking steps to reduce its weak position against private companies. These measures have improved social conditions to an extent, and also they have the potential to stabilise informal work. However, legalists claim they result in exclusion of members of the informal economy from access the formal system. There is certainly evidence of this occurring in Buenos Aires, as Green Points are unlikely to absorb more than a small percentage of urban recoverers.

It appears that the city government is struggling to strike a balance between increased regulation, which allows for greater social protection and reduced economic vulnerability for some, and decreased regulation which leaves informal workers vulnerable to the forces of the free market but enables the absorption of a huge workforce, unrestricted by governmental red tape.

It is this enduring conceptual debate that, on the ground, has resulted in the development of other forms of structural support where government policy has either been inappropriate or ineffective. The cooperative movement appears to have great potential for providing a larger degree of organisation and formalisation to urban recoverers, through greater bargaining power with private companies, increased social support from other cooperative members and a new form of organisation by which they can be officially recognised.

Another basis for the development of grassroots support structures is the unpredictable nature of public opinion and, therefore, government policy. Chronopoulos, T. (2006) raises concerns over what could happen to informal workers if the then current Chief of government, Ibarra's sympathetic approach was not continued by future governments. In December 2007, Mauricio Macri's 'Commitment to Change' Party was elected into the municipal government. Given Macri's previous statements regarding the removal of informal recoverers from the city's streets, his personal financial interests in the system and continuing focus on unrealistic types of formalisation, there is just cause for concern. Increasingly marginalised, informal workers will be allowed or encouraged to 'disappear' as they lose their place in the system that sustained them throughout a range of political and economic circumstances, and not provided with alternative options. Constant research, monitoring and adaptive policies are required if appropriate actions are to be taken, because one thing above all others in this situation is evident: both the hand of the government and the free market have ensured that the future of urban recovery is vulnerable and uncertain.

This chapter presents a case study of the role of the government in Buenos Aires and the policies which influence the urban recovers WFL would be working with. It is clear that any way in which the vulnerability and fragility of the urban recoverers can be lessened and any way in which their economic stability and independence from the changing and potentially hazardous government policies can be enhanced will be of benefit to these groups and is worth pursuing.

EXERCISE

Understanding the Context. Stakeholder focus.

Hold a role play debate. Prepare your characters beforehand and come ready to defend your position on the points below.
Group 1. The local government
Group 2. Independent cooperatives of cartoneros

1. Should Green Points (sorting centres run by co-ops) be created?

2. How should responsibility for recycling be divided amongst the stakeholders?

3. What should happen to the street cartoneros?

CHAPTER 6

Stakeholder Focus: Cooperatives

After initial exploration of the context and in relation to the details regarding the lives of the cartoneros in BsAs as given in Chapter 5, we decided to work with cooperatives of cartoneros – those collectives who were already organised and ready to move from collecting and sorting to processing and manufacturing. In this chapter, we look at some of the issues we needed to understand and consider before deciding who we would take the next steps with.

One of the main criticisms for being involved in work of this nature is well founded. The development agenda, as Ferguson points out, has become a machine that implements technical solutions to problems that are far from technical in nature (Ferguson, J., 1990). Technical knowledge is often confined to that of the 'expert,' and hence it is easy to create a power structure and hierarchy when it comes to technological development. Well negotiated, participatory socio-eco-technical solutions are hard to find. Despite much recent attention to 'public dialogue' with science, in our opinion, this is not more than governments wanting to suppress the complaints of the public so that innovations can reach the market more quickly. It is often stated that the 'lay' public does not know enough science to be able to contribute to the discussion (Sclove, 1995). There are, however, several groups, which actively try to create programs for communities and citizens that are responsive to technical needs analysis (Sclove, 1995). Coburn raises the contentious political questions that professional 'techno' science tries to silence by often claiming that an issue is 'purely technical.' Gupta, A. (1999) tells us 'formal scientists do not recognize, respect and reciprocate the informal scientific knowledge, creativity and innovation at grassroots levels in society. The science underlying the successful overcoming of some of the day-to-day struggles of economically poor but knowledge rich people does not get articulated or acknowledged...' (Gupta, A., 1999, p. 368). His solution to move beyond the barriers, blending excellence in formal and informal sectors is the 'honeybee network' (http://www.sristi.org/honeybee.html) in which 75 countries draw on technological and institutional innovations for sustainable natural resource management developed by people unaided by NGOs, market, or state. In BsAs, they call this 'socializing knowledge,' and it works on a much more local level. One of our first encounters with cooperatives was encouraging – we were told, 'you share the knowledge with us and we will share it with others.' They had the 'scientific' knowledge about what could and could not be recycled locally and what processing methods were realistic and manageable. We had the knowledge of how to make the composites. The challenge was to work in this shared, non-exploitative, counter hegemonic way with the cooperatives in BsAs.

A key aspect of co-creation and co-location of projects and programs is the mutual respect and recognition of values. This is the most difficult area to translate/transform when working with students. Do we work within our value system or that of the communities we work with? What if we don't agree with their values or they with ours? What if the values of the groups we work with clash with each other? Within WFL, we choose not to work with groups whose values differ substantively from ours because it is not our position to change the values of communities we work with. Values underpin the work of WFL and its focus on social and ecological justice, gender equity, and poverty reduction.

What reverberated for us in BsAs, and what was significant for WFL, which at its core depends upon open and collaborative knowledge building and dissemination, was this practice of 'socializing or collectivizing knowledge.' It is a starkly different ethos than that held by small companies we see in North American and the UK, which have been taught to say, 'don't tell the others or we won't be able to sell so much.' Our work with the coops showed us that different groups work within the system in different ways.

In the section below, we present briefly each group that we encountered, to see if they were interested in working with us, or us for them in ways which would enhance the cooperative spirit and not diminish it in any way.

6.1 THE COOPERATIVES

6.1.1 EL CEIBO

To the world outside of Argentina, El Ceibo, which is made up of 50+ families, is the most recognizable cartonero cooperative. It has been studied by university researchers and featured in the popular press. Cristina Lescano, its founder, has more than 20 years experience as a successful community organiser, and she was the very first person we met in Buenos Aires. We didn't have much to offer Christina in July 2007. Waste for Life was simply an idea. We had no hotpress in Argentina; we had no product; we had no idea whether it would be economically advantageous for her group to spend time collecting and sorting the plastic bags and cardboard – time spent away from doing other work. El Ceibo is renowned because of its successful collaboration with residents of the Palermo neighbourhood in BsAs who sort their own garbage and make the recyclables available to the cooperative. Christina would work with us when we were more certain that our technology could be turned into a commercial success. Until then, she suggested, we should spend time with Carlos Levinton's researchers at the University of Buenos Aires.

6.1.2 BAJO FLORES

The Bajo Flores Ecological Cooperative of Recyclers was the first cooperative to benefit from the Zero Garbage law and, during our 2007 visit to Argentina, the only 'green point' in operation. The Buenos Aires city government built the plant, bought the machinery, and loaned the installations for five years to the members of the cooperative. Bajo Flores is supposed to receive the raw garbage of 5-star hotels and over-19-story apartment buildings from the exclusive Puerto Madero district of Buenos Aires. Two of the five private trucking companies and the single government-run company that are responsible for cleaning the streets and hauling the city's garbage to the CEAMSE landfill

Figure 6.1: Bajo Flores Cooperative.

were charged with bringing the waste to the cooperative to be sorted. The cooperative members separate the different types of recyclable plastic, bottles, carton, paper, etc., from the soft garbage – which is then hauled to the CEAMSE landfill – and then sort and sell it to the industries that recycle them into paper and other plastic and glass products. In theory, it's a good system: the cooperative is responsible for sorting and selling, but not collecting, and it bypasses all of the middlemen that stood between its members and the final destination of the recyclables when they were working informally as cartoneros on the streets. The members share the profits more or less equally, less garbage goes to the landfill, and the plastic and bottles and cardboard are reused. The city government, for the price of a warehouse, two pieces of equipment, and a small subsidy that goes to cooperative members gets much waste processed – a small price to pay to keep the peace. But during our 2007 visits to Bajo Flores, it was obvious that the Green Point was working way below capacity. What was happening to the garbage along the route? Though destined to Bajo Flores and the 40+ families working there, we learned that the truckers or the hotel employees or the apartment building janitors were diverting it, and someone was selling the recyclables on their own. There are lots of competitors for garbage and many people could profit making certain that Bajo Flores failed. By the time of our return visit in 2008, Bajo Flores had fallen out of favour with the current government and was barely functioning, and the Director has been accused of (unconfirmed) fraudulent dealings.

6.1.3   VILLA ANGELICA AND ETILPLAST

Two cooperatives funded by The Working World, Etilplast and Villa Angelica, are examples of small family coop organisations that buy waste plastic and reprocess it to add value to the chain. Villa

Figure 6.2: Etilplast Cooperative.

Angelica has a yard full of waste and two machines that chop up the plastic. When we visited, we were told that they had recently lost their only client – their truck and small extruder had broken down, and they couldn't seek new buyers until they were mended. But they didn't have the funds to do so. All of their equipment was priced in terms of the number of tons of plastic waste they had traded to get them, and these machines were worth less than a few thousand dollars. Villa Angelica was in too fragile a position to even consider working with us. A more successful organisation, Etilplast, had an enormous industrial extruder, which they had built themselves over some eight months and

Figure 6.3: Etilplast Cooperative members.

which enabled them top process an intermediate product – plastic pellets. They were comparatively thriving and were interested in collaborating with WFL, recognizing that the combination of their extruder with the hotpress could very effectively manufacture products.

6.1.4   AVELLENADA

Carlos Perini and his cooperative Avellenada, south of BsAs, have moved through several stages since we met them in 2007. They appeared very disorganised and dysfunctional on our first visit, but a few months later had organised a cooperative to collect, hand sort, chop, wash and dry plastic for sale by type. They suffered, however, as have all of the cooperatives and all of the street cartoneros from the precipitous decline in the value of recyclables due to the 2008 world economic crisis, and by the time we returned, they had closed this part of their business to focus on a scrap wood furniture making sideline. While previously, they were able to support 50 cartonero families, this is now down to 25. They have a waste wood chip product that would work well as reinforcement fibre for the composites, and they already have the furniture making skills to consider a range of products.

Figure 6.4: Villa Angelica 1.

6.1.5   RECICLANDO SUEÑOS

Reciclando Sueños is the only coop we know that has moved to manufacturing from collecting, sorting and selling. They have a product – a painting sponge – the handle and backbone of which is made from recycled plastic. They chop, clean, dry and injection mould the plastic and then assemble the parts, selling them each to a local wholesaler for 1.10 pesos (29c US). This cooperative has an interesting position in relation to the municipal government in Matanza where they are located. They consider their work a public service that the government has failed to provide, and see no sense in the current arrangement of paying large companies to collect waste. They have refused up to now to accept (or rather apply for) any form of government welfare checks or work plans, insisting that their work is legitimate, and they should be paid by the municipality for the services they provide. In 2007, Reciclando Sueños was well positioned to move forward with WFL, but in December of that year their warehouse burned down. The problem is that plastics are very flammable and take up much storage space and get placed in the same warehouse as faulty electronic equipment. They have since been recuperating from their losses.

Figure 6.5: Villa Angelica 2.

6.1.6   COOPERATIVA EL ÁLAMO

In 2007, El Álamo was working out of a damp pit at the side of a railway track in Puerreydón on the outskirts of Buenos Aires. It had been designated to manage one of the city's 'green points,' but the selected location triggered a powerful NIMBY (not-in-my-backyard) backlash. It was unclear how we could work with El Álamo because they concentrated on collecting only cardboard. By the time of our return in 2008, they had relocated to brand new warehouse facilities adjacent to Bajo Flores. The cooperative was prospering and was clearly benefiting from the largesse of the current Buenos Aires City government. New trucks were moving in and out of their facilities all of the time, and cooperative members were clothed in uniforms blazoned with their logo. This was in stark contrast to its neighbour, Bajo Flores, which had become only a shell of its former self.

6.1.7   ABUELA NATURALEZA

Maria Virginia Pimentel of Abuela Naturaleza has decided to spend her life working with waste recycling. Formerly, a cartonero organiser, she has been involved in recycling efforts since the 1980's

Figure 6.6: Reciclando Sueños Cooperative members.

and now teaches children about recycling in schools and fetes using puppets and other creative techniques. She insists on being called an 'urban recoverer' and has organised her local community to separate products that can be recycled, which she collects in a van. Virginia has found a market for practically everything she collects and showed us a huge number of items that she had sorted with the help of the three other cooperative members. Maria Virginia conceived of a recycling box made of waste as a potential WFL product, for her neighbours to use to make separation easier for them (no 'blue boxes' are provided by the government). She goes once a week to the Faculty of Architecture, Design, and Urbanism (FADU) at University of Buenos Aires to visit our partner Carlos Levinton who runs the Centre for Experimental Practice there, to work with the hotpress and develop product ideas and experiment with different materials.

Figure 6.7: Reciclando Sueños paint sponge.

6.2   CIVIC ASSOCIATIONS

6.2.1   RENACER LANZONE

Renacer Lanzone or the Pheonix of the barrio Lanzone is not in fact a cooperative but a civic association whose president is Adam Guevara. Over twenty years ago, Guevara had the idea that CEAMSE should divert the trucks carrying garbage to its landfills to groups of cartoneros who could strip them of their recyclables, leaving less waste to eventually bury. He tried to persuade many CEAMSE directors to adopt this idea, but was unsuccessful until three years ago when the current director agreed. Adam was allowed to establish a model 'social factory' or sorting centre in the Lanzone barrio across from the Norte III landfill. The project has been so successful that there are now four centres and four more planned on the CEAMSE site itself. It is difficult to determine how independent Guevara's group is from CEAMSE at this point, but they are very keen to work with WFL.

Figure 6.8: Installing a hotpress.

6.2.2   RECOVERED FACTORIES

These organisations are in themselves a fascinating social phenomena. Many businesses went bankrupt in the economic crisis of 2001 and factories lay inoperative. Several factory workers 'took' over the factory and by a quirk of the law became the legal operators of their own plant. Many still continue to be run as equal pay co-ops, some eight years later. These are known as recovered factories.

6.2.3   UST

UST (Union Solidaria de Trabajadores) in Villa Dominico is a cooperative and 'recovered enterprise' that extracted itself from the withdrawing corporation TECHINT and successfully negotiated with CEAMSE to become contract workers who would maintain regenerated land on top of a large former landfill. Their expertise is sanitary engineering, but they work alongside and support other cooperatives that work in a variety of areas including recycling and building. UST says it has the skills to build a hotpress.

Figure 6.9: UST Cooperative members.

6.2.4   19 DE DICIEMBRE

Cooperativa de trabajo '19 de diciembre' is a recovered metallurgical factory. It continues, as before, to manufacture auto components for the automobile industry and, ironically, one of its principle clients is it's former owner who has created another factory which buys and assembles the components and sells them on to the large automobile manufacturers (Ford, VW, Mercedes Benz). The '19 de diciembre' employs about 5 or 6 outside skilled workers who work the more complex die machinery and are paid $4/hour, which is twice as much as the 30 or so cooperative members earn. (It is typical for skilled workers to either leave or not join cooperatives because they can make more money as 'independent' employees.) It is important for us to find a local 'recuperated' factory that is able to manufacture and sell the hotpress as a niche business, and Enrique Iriarte, the Cooperative's president expressed interest in doing so.

6.3   ASPECTS OF COOPERATIVISM

Our work with the co-ops showed us that different groups work within the system in different ways. All fight for the right to work, to own their own time, to not labour for others but to share the proceeds equitably of the work they contribute. They have – they have been forced – to reorganise their working lives, to provide for themselves and their families. Some important features of this include:

ASSEMBLIES AND EQUAL PAY

Many co-ops function as the recovered factories do, having weekly membership assemblies or workshops in which major decisions are made and issues of 'being a cooperative' (which are neither evident nor easy) are collectively worked through. Reciclando Sueños, for example, has two anthropology researchers help facilitate these meetings. We had the privilege of attending one of these assemblies where members discussed, among other things, what should be done if one of them collected a pair of socks that was still wearable. Should they bring that pair back to the coop or keep it for themselves? It was decided they would bring the socks back, eventually sell them, and share the proceeds – though none of them were wearing socks that winter day.

SIN PATRÓN

Despite the fact that all the coops we worked with had a designated leader, this person did not have any power over wages, conditions of work, or any other important considerations. All decisions are made together in a non-hierarchical, participatory, decision-making process, and in most cases, pay is equally distributed. Although we heard this phrase ('without a boss' or 'without an overseer') most often from independent cartoneros who used it to differentiate, positively, their lives from those subject to the collective decision-makings of a cooperative, sin patrón gained currency as an idea during the recovered factory/self-management movement that began with the 2001 crisis and is grounded in the ideas of cooperativism.

PEOPLE BEFORE PROFIT

One of the clearer forms of resistance we encountered was how people were brought into the coops. In most cases, if they needed to be included in group or family cooperatives, they were included. Civic Associations have a different structure – they are not legally cooperatives – and they do have a president and wages. However, Renacer Lanzone adhered to the same principles as the cooperatives. When asked on what basis he selects people to work in his association, Adam Guevara told us 'on the basis of need of course.' If an unemployed woman has five children and another, three, then the first one gets the job. This is clearly unskilled labour in the main, but it is a far cry from the merit competitive systems we are accustomed to in market-driven economies.

SELF-SUFFICIENCY

Faced with no jobs, no social welfare, and plenty of garbage, these individuals and families have grouped together to become self sufficient and eke out ways to survive. The coops have found many imaginative ways of gaining more income. Renacer Lanzone sought out education, and were trained by representatives from the Instituto Nacional de Tecnología Industrial (INTI) to categorize and sort different plastics, which increased their income by 300%; El Ceibo, Abuela Naturaleza, and Reciclando Sueños have organised neighbourhoods to separate waste before their members collect it; Etilplast and Villa Anjelica process much of the plastic they collect, which they sell at higher prices than the raw material alone; and Reciclando Sueños has ventured into small manufacturing. Many saw our intervention as another way of contributing to their self-sufficiency, of helping them resist so they could continue to explore their alternative modes of social and productive life.

EXERCISE

An Ideal Final Result

Create an ideal final result plot (IFR) for each cooperative / organisation discussed above. IFR plots can be developed whereby the items on the outside represent the ideal scenario and the pointers mark the reality of the situation from zero at the centre, so you can see which need attention. Only those items desired by the co-op should be included. Usually this would be carried out by the co-op themselves rather than by you guessing – but it serves to demonstrate the differences between each group. An example plot follows (you can add many more items of course).

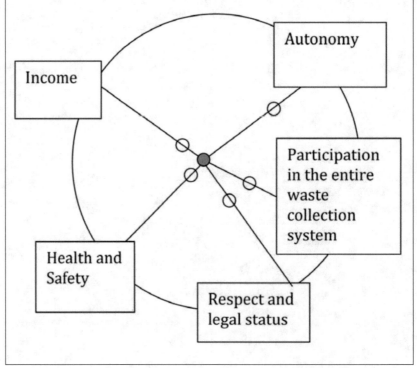

CHAPTER 7

Sustainability: Economic, Environmental, and Social

One of the most important factors to consider with regard to any form of intervention in a community, particularly amongst more vulnerable populations, is not to create false hopes about the outcomes of a project, nor to abandon that community mid-flow, Likewise, it is important to ensure that there will be no negative and unexpected impacts, one of the most harmful of which is the creation of a state of dependency on external partners. During the first phases of the WFL BsAs mapping, we were constantly questioning what the project would look like beyond our participation. How could it be self-sustaining? It is important to build into any project an evaluation scheme and monitoring that continually questions the initial goals, progress, and is a predictor of eventual impacts. It is not enough for scientists and engineers to claim that they could not predict the usage of their technologies. They need to anticipate as far as possible into the future in consultation with potential users, all possible ways in which the technology might be used. They can then work with the community they are developing the technology with to create solutions for safe and healthy adoption of the new system, with positive social, environmental and economic impacts. Although this will only ever be an approximate art, it is as accurate and as critical as a numeric calculation, which approximates for unknown parameters and builds in error margins. Ideally, we should balance economic, social and environmental factors for the groups we are focussing on. However, often the economic and the environmental favoured to the detriment of a deep analysis of social impacts. For instance, Greenpeace might favour recycling however it is done – perhaps as efficiently as possible in a large commercial factory, but the affect on the cartoneros could be economically devastating. We look below at Ursula Franklin's triple bottom line and map them onto a scenario in which the cooperatives, Green Points, and social factories create products from waste. In the space available, we can only begin to touch on the sorts of issues involved, and we hope that it will urge the reader to further reading in each area.

7.1 ECONOMIC SUSTAINABILITY

In the economic analysis for Waste for Life, we need to consider the available market for an eventual product or products. Assuming that there is a good market for the product, we then need to look at costs of materials production and compare them with current income levels. As discussed in Chapter 2, we must know if there is there a steady supply of waste plastic and fibre and know what alternatives there might be and at what costs if the original source dries up. Additionally, we have to concern ourselves with ancillary costs of production e.g., electricity and the technical solutions to lower energy costs (also feeding into environmental sustainability). In Table 7.1, a very simple cost benefit analysis has been created in which we can see at a glance the potential income for various co-ops considering a variety of materials and products, which gives us an idea of the economic feasibility of WFL. Additionally, when variables change, such as the cost of electricity or the market price for selling recyclables directly to agents, the table can be used to recalculate and continually monitor the projects economic viability. The example below has been developed for production of a ceiling tile and prices are in pesos. It is simply comparing selling the processed product versus the collected materials directly and does not take into account labour. It is assumed that the co-op members would work equal hours and divide profits.

Table 7.1: Requirements.

Material requirements and costs

The production of one tile requires 1.661kg plastic and 0.704 kg cardboard. The total amount of raw materials required depends on the production target. The following calculation assumes a weekly production of 50 tiles. This number can be changed in this spreadsheet.

Calculation Step 1 (required material per tile * weekly production target)

Plastic required per week:	**83.05 Kg**
Cardboard required per week:	**35.2 Kg**

In the following the relative surplus that could be achieved by producing ceiling tiles rather than selling the raw materials is estimated based on the revenue received from selling the raw materials versus the projected income from selling ceiling tiles. If the price per kg plastic or cardboard varies it can be edited (All prices are in pesos)

$1.50	**Price of 1 kg plastic**
$0.28	**Price of 1 kg cardboard**

Clearly, this is a very rudimentary analysis, but it works as a good first step in evaluating whether it's worthwhile taking further steps. The cooperatives would be encouraged to do a full scale business analysis if they were to borrow money from the hotpress loan scheme of Working World or any other microcredit organisation. In this case, which assumes a good enough market,

Table 7.2: Costs.

Equipment Costs

Hotpress:	**$2,375.00**
Annual interest rate:	**10.00%**
Amortisation period (years):	**5**
Monthly capital costs:	**$50.46**

Operating Costs

Repairs and maintenance per year (estimated):	**$1,100.00**
Electricity required per tile (kwh):	**0.3**
Electricity costs ($/kwh):	**$0.16**

Total Costs

Monthly costs	**$721.90**
Annual costs:	**$8,662.85**
Costs over 5 year period:	**$43,314.24**

Table 7.3: Revenues.

Revenues

The revenues in addition to selling the raw materials at market prices strongly depend on the retail price of the ceiling tiles. The currently estimated price can be edited.

$7.91   **Estimated product price**

Monthly revenues:	**$1,718.54**
Annual revenues:	**$20,622.50**
Revenues over project period:	**$103,112.50**

Additional Profit

This is the projected profit in addition to the surplus that could be achieved by simply selling the raw materials.

Monthly Profit:	**$996.64**
Annual Profit:	**$11,959.65**
Total Profit:	**$59,798.26**

it can be seen that producing ceiling tiles from the collected raw materials can be a value adding process to the co-ops, so it is deemed worth of further exploration. Other products can be assessed in a similar way but what is also required is to assess the potential market available for the various products. The next section looks at aspects of the marketplace in Buenos Aires that might affect the sale of potential WFL products and which form the basis of information on which product choices might be made. It is worth noting how different these issues are compared with a less developed country such as Lesotho and how economic and marketing tools must be adapted to ensure that our stakeholders are the key beneficiaries and that social justice comes first.

7.2 FACTORS AFFECTING POTENTIAL SELLING POWER OF CARTONEROS ADOPTING WFL TECHNOLOGY

7.2.1 COST OF LIVING OF ARGENTINIANS

The cost of living in Argentina is surrounded by feelings of insecurity and anxiety by slowly recovering individuals. In 2003, Néstor Kirchner, ex-president, received a devastated economy with an unemployment rate of 21%. In 2006, the unemployment rate lowered to a 10,4%. To manage inflation, the government implemented methods of regulation negotiating prize accords with supermarkets, pharmacies, and meat producers. In 2005, the 'cost of living' reached 12.3%. Poverty still affected a fourth of the 37 million argentines (Garcia, M., 2007).

7.2.2 INCOME

Family income is the next step to defining where Argentines stand within their economy and the amount of money they can dispense. The lower upper class family income is above $3,060, while the higher upper class have a family income of $5,700 and higher. The 40% of the population in the lower class have a family income of $740, while the lower middle-class has a family income of $1,325. Due to the variety of incomes developing, the bridge between the upper 10% and the lower 10% is widening (Bullentini, A., 2008).

7.2.3 BUYING POWER

Compared to all the countries in South America, Argentina's citizens have the largest buying power. It is estimated that with 1,400 pesos, an Argentine worker can buy the same goods as an American worker can with USD 676. Those who receive minimum salary are considered "poor today because of today." There are nine sectors where salaries have increased more than average: agriculture, cattle, hunting, reforestation, mine exploding and quarry, electricity and water, social and health services, community services, social and personal, domestic service, transport, storage, and communications. Salaries have been reduced in real-estate, commerce, public and defense administration, financial mediation, manufacturing industry, hotels and restaurants, non-resident employees, and fishing (Fabiani, A., 2005).

7.2.4   DISPOSABLE INCOME ISSUES

Argentinean consumers, along with others around the world have felt the bitter after effects of the dwindling global economy. As a result, consumers have become more susceptible than before to defend their income and spending which is where the need for justifiably high quality, value added goods comes in. Companies have to add value through better attention, with a better service or with a proposal that addresses consumer needs.

7.2.5   FRAGMENTED VS. NECESSARY GOODS

Another important factor to determine the buying power within a society is the differentiation between fragmented and necessary goods. Fragmented goods are products that can only be afforded by specific classes, whereas, necessary goods are products that are bought and afforded by all classes. In Argentina, for example, telephone land lines are a service that most of the population can afford and pay for. Home internet access, on the other hand, is a service that only the upper and middle classes can afford or invest money in. The fragmented vs. necessary goods statistics also demonstrate the priorities of the average consumer. In Argentina, the service with the most money invested in is the telephone land line service. From this statistic, it can be assumed that most Argentine families find it important to have a land line in their home, as it may be their main form of communication. After this, the next largest investments consumers are willing to make are cell-phones, which are followed by medical insurance. Not only does this information explain the priorities of where Argentineans invest their money in, but it also shows what services are out of reach for lower classes due to economic instability.

7.2.6   DISTRIBUTION

Types of intermediaries

- Wholesale. (Wholesale) Wholesale trade is an intermediary that is characterized by selling to retailers, other wholesalers or manufacturers, but never to the consumer or end user. Wholesalers may buy from a producer or manufacturer and other wholesalers.

- Retailer or retailer. (Retail) Retailers and retailers are selling products to final consumers. Retailers are the last link of the distribution channel, which is in contact with the market. They are important because they can change by retarding or enhancing the marketing and merchandising activities of manufacturers and wholesalers. They are able to influence the final results of sales and marketing articles.

7.2.7   CHANNELS OF DISTRIBUTION

- Direct Channel. The producer or manufacturer sells the product or service directly to consumers without intermediaries.

- Indirect channel. A distribution channel is often indirect, because there are intermediaries between the supplier and the end user or consumer.

- A short channel has only two steps, i.e., a single intermediary between the manufacturer and end user.

- A long channel involves many middlemen (wholesalers, distributors, wholesalers, resellers, retailers, dealers, etc.).

7.2.8   CHOOSING DISTRIBUTION METHODS

Once you have selected and developed a unique product or business idea, correctly positioned and targeted it to buyers, and developed your packaging and pricing, the selection of distribution channels and sales representation is key to successful marketing. In the case of a general unstable economy and political environment, the different faces of the product, the pricing, packaging, and product itself enables would ideally be versatile and adaptable to prevailing market trends. However, distribution and sales decisions, once made, are much more difficult to change. And distribution affects the selection and utilization of all other marketing tools.

7.2.9   PRIORITIZING DISTRIBUTION OPTIONS

Small businesses can employ several different channels for the distribution of a product. However, most small businesses must prioritize distribution channels and sales force options over the evolution of the company, e.g.,

- Multi-level "network" organisations, with company and independent sales reps.

- Department stores, with company reps and sales brokers.

- Mass merchandise stores, with company reps and sales brokers.

- Club member warehouse stores, with company reps and sales brokers.

- Direct mail, with company personnel.

- Distributors, with company sales managers, brokers, distributor sales reps.

It is not always possible for a company, small or large, to take advantage of all possible channels that match the marketing strategy it wants to achieve. Financial considerations aside, it is important to prioritize and utilize each distribution channel to suit the needs of a particular product rightly taking into the context the considered markets and prospective competition.

7.2.10 TOURISM PROFITS IN 2009

The Consultant Argentine Agency of tourism (CAT), a non-profit organisation whose main purpose is to "represent, defend, and promote the development of tourism service in Argentina, estimated a downfall in tourists profits for this year. Due to the international crisis, it is estimated that tourist profits will diminish by 7% this year. This year's tourist profits estimated to be USD 4,000 million against the USD 4,300 in 2008. It will be crucial to recover the 2007-2008 numbers in 2010. For now, the current receptive tourism will continue to grow at a slow rate, meaning that the country will still be strongly dependant on inner tourism (iEco, 2009).

7.2.11 POTENTIAL MARKET FOR GOODS

Through this research, it can be deduced that the kind of products we are proposing for the cartoneros to manufacture have to cater to individuals in the upper and middle classes. The current cycle time of the hotpress equipment makes the production of most products unfeasible. The cartoneros will not be able to compete in the market place producing consumable or low value products. The products produced will need to have value imbedded in them in other ways. This can be achieved through a couple of different avenues. These can include the design of the product, craftsmanship or inherent cultural relation. With these considerations, the relative cost of the products produced will need to be higher in order to be sustainable. Regardless of the market that the products are being sold in, the products produced will be purchased by consumers with disposable income as opposed to targeting necessities that would apply across a larger consumer base. Based on current conditions and the unreliability of skilled labor at this point, products for initial production should not rely heavily on specialized equipment or labor. This will present many design challenges in regards to the manufacturing of the products, but it will enable a justifiable and viable starting point for the cartoneros to start from.

Keeping this in mind, a product manufactured by cartoneros or an organisation of cartoneros, distributors and vendors would have to make sure that the average consumer reflecting the majority of the population is comfortable with purchasing that product. Argentines are greatly affected by their perspective on their social and economical status. Therefore, it is important to show them that despite belonging to a "recovering middle class," they have the ability to purchase our product because it is made for them. It is made for them because it takes into account where they are coming from, their economical troubles, and aesthetic opinions. From this communication, we can then show them that our product was made to qualify for their lifestyle and needs, and that our price range takes into account their spending abilities. We also understand that even though our clients may feel that they are not economically strong, they want to receive a product that is presented to them in a manner and environment that treats them as if they were well off financially.

Products catering to tourism allow for a cultural relation and hence value being added to the product. This avenue would allow for higher margins while giving a cultural relationship for the cartoneros as well as the consumer. Tourist products in Argentina reflect high craftsmanship and cultural motifs. They cater towards individuals in search of beautiful crafts in leather, silver, wood, and other materials that enrich their lives through style, history and originality.

If and when a product proposed for manufacturing by cartoneros is placed for distributing within the tourism industry, we can assume that the product should not replicate tourist products that exist in Argentina because plastic is not a traditional material in Argentine craft. Plastic may be able to produce similar items, yet, the essence of Argentine taste will not be in the items, creating a possible downfall. These products will need to reinterpret or create a new line of product that reflect the cultural heritage of Argentina.

7.3   ENVIRONMENTAL SUSTAINABILITY

Often, materials and businesses call themselves 'green' in marketing, but they would not be able to prove this status to anyone. The reason they can get away with this is because it is not easy to determine if the product or process you are developing is really beneficial or at least less harmful to the environment than alternatives. Even the available tools to help make such an assessment have their limitations. They can be inaccurate and take much time to complete with any degree of effectiveness. However, it is possible to get a good comparative evaluation between two variables with relative ease, and it is feasible to locate other studies which have been done on similar products and make some judgements based on these.

Lets look at some common misconceptions and assumptions made:

1. If we use recyclable packaging our product is 'green.'

   Not necessarily. Packaging is generally a bad idea and should be avoided. Recycling and reprocessing can sometimes be worse than putting products into landfill, if we take into account the energy used for the process and pollutants emitted into the atmosphere.

2. If we use naturally available plant materials to reinforce a composite material, it will always be 'green.'

   Not necessarily. The plant may take a lot of energy to grow, it may be genetically modified and require pesticides to fully develop. Furthermore, in balancing against social factors, it may compete with the plant or other crop being used as a food source in the local area.

3. Using waste plastic in a product usually made from plastic makes it 'green.'

   Usually this is the case, but we must assess whether the energy used to recycle the material does not overrule the advantage of removing it from landfill.

One of the best ways of getting a handle on whether something may truly be called 'green,' e.g., that the environmental impact is reduced compared with previous or competitive products or contexts is to conduct a Life Cycle Assessment or LCA. Again, this will not give an accurate number or a yes/no, but it will be able to provide some evidence for comparative studies. One of the major disadvantages of the selection of data. Datasets are still being created and ones that do exist are owned by companies who either charge large sums for its use and do not divulge sources/methods or do not share their data at all.

Life cycle assessment (LCA) is a tool used to evaluate environmental impacts associated with a particular product throughout its entire life cycle (from extraction of raw materials to the end of life, from cradle to grave) [(Murphy, R., 2003) and (SETAC, 1993)]. LCA can be used to compare two or more products to find which of them is more preferable from an environmental point of view. For a fair comparison, products should meet the same service requirements. Also, since different products have different impacts at different stages in their life times, comparing them at only one stage can give misleading results. Their whole life cycles should be considered. The automotive industry, for example, makes a significant contribution to environmental pollution, especially in emitting greenhouse gases. But researchers in the field of natural fibre composites or NFCs suggest that these products can help reduce the harmful environmental impact of automotive materials (Li and Wolcott, 2004). Wotzel et al. (1999) and Diener and Siehler (1999) as reviewed by Joshi et al. (2004) indicate that NFCs are more eco-friendly than glass fibre composites in automotive applications because (1) natural fibre production has lower impacts than glass fibre production: they depend mainly on solar energy to grow and they absorb CO_2 during growth (2) natural fibres take more volume per unit weight of a composite; therefore, NFCs require less polymer matrix (which has more impacts on the environment) than glass fibre, composites, (3) due to their light weight, NFCs improve automotive fuel economy, thereby reducing emissions during use (Sivertsen et al., 2003).

LCAs normally begin with goal and scope definition. This is the stage that defines the goals of an LCA and delineates boundaries of the study. It determines the outcome of the LCA. An LCA goal must be stated precisely to leave no room for ambiguity (UNEP, 1996). For most studies, goals may fall within the following areas: comparisons between two products, product and process development, decisions on buying, structuring and building up information, eco-labeling, environmental product declarations and decisions on regulations (UNEP, 1996). One may ask specific questions to formulate goals. Will the study be for internal or external applications? Who will form the audience for the results of this study? To what level of complexity shall the LCA be confined (Murphy, R., 2003)?

Then system boundaries determine the extent of assessment. LCA practitioners decide which inputs, outputs and processes to include or exclude (Murphy, R., 2003). Where will boundaries be drawn and how? Gokoop and Oele (2004, p. 5) illustrate, 'In an LCA of milk cartoons, trucks are used…to produce trucks, steel is needed, to produce steel coal is needed, to produce coal trucks are needed…one cannot trace all inputs and outputs….' Undoubtedly, the question of where to draw a line in this endless chain is a problematic one.

Our own studies using LCA (Thamae and Baillie, 2008) of natural fibre composites made from wood fibre indicate significant reduction in life cycle impact compared with the use of glass fibre. In that study, as discussed above, certain assumptions have to be made and data sets selected. As composites are relatively uncommon materials compared with other categories such as glass, cement, metal, plastic and as LCA datasets are still young in their development, we had to make some next-best choices for data which seem rather strange. For example, there was no available category of glass fibre so we had to use data for glass. Furthermore, cardboard fibre would not exist as a category, so we used wood fibre instead. In the study, we compared different kinds of plastic which could be used and verified that thermoplastics are much better than thermosets as expected. Thermosets are typically resins which 'set' and cannot be later recycled. Although composites are not the best for recycling with any plastic base or 'matrix' as we call it, at least if the fibre and plastic can be remixed and chopped and melted, some form of lower quality product may be created. Glass is bad for recycling as it tends to cut up the reprocessing equipment, and it is possible to recover and use energy by incinerating natural fibre but not glass fibre. Any form of thermoplastic or plastic, which can be melted, is therefore better. However, if the material gives off fumes (such as PVC), this can act against the positive impact. Finally, if waste is used, we can balance this against the effects of the materials going to landfill, which would always be preferred. Hence, our overall conclusion is that compared with any form of 'artificial wood,' fibre board made with resin, or glass fibre reinforced plastic, products made from WFL materials would have a lower environmental impact. This would be especially true of any product which would use fuel due to the weight reduction of these materials compared with traditional materials. Once a product is identified, a full LCA could be enacted as required.

7.4   SOCIAL SUSTAINABILITY

Social sustainability is not as well studied, understood nor as well developed as the other two pillars and is clearly difficult to identify and to measure. It includes the cultural, social and political issues, which lead to a community and an individual's well being, and freedom to be a full citizen in a society with a safe, healthy life, enough to eat, drink and sustain oneself through employment or other legal means, and access to education and healthcare (Sen, A., 1999). Waste for Life has identified the following social factors which success and sustainability might be measured against. The ultimate measure is the numbers of individuals/cooperatives with extra income generated per month. Although this is an economic factor, the social aspects that accrue to this are associated with the freedom that economic independence allows – meeting basic needs, e.g., not only food to eat, but the choice of what food to eat. There is of course a blurred boundary in between all of these categories, the only reason we separate them is so that we don't forget any of them.

The success of WFL BsAs/Argentina and the ability of the cooperatives to sustain it, in addition to their current businesses of collection, separation, storage and marketing of recyclables, is dependent on the following being in place:

1. Ability and knowledge (e.g., training) of cooperatives to maintain the WFL business – collection, separating, production

2. Local, made to order production of the hotpress machines (from local materials) and any other equipment as required, plus the funds to run and the ability to maintain the equipment

3. Access to local microcredit organisation/cooperative one-off funds to support the purchase of the hotpress

4. Independence from local government and business as well as ultimately UBA, WFL and local NGOs, as well as other university partners

5. Balance – homeostasis with broader range of actors, e.g., in BsAs, CEAMSE, social factories, street cartoneros

6. Healthy and safe working environment – e.g., protection from fires, poisonous fumes from burning plastic, etc.

7. Safe products – e.g., building and product codes

8. Promotion of equitable communities, does not depend on or exacerbate power relations

9. Enhancement of gender equity and respect of child labour laws

10. Technical solutions to the reduction of maintenance costs, e.g., electricity

11. Sustenance of flow of recyclable materials. Continual assessment of changing prices of recyclables. In Lesotho, agricultural sustainability of Agave plants

EXERCISE

Who pays and who benefits? Who or what are we sustaining?

For this exercise, in order to understand more fully the kind of decisions which need to be made by engineers, we will adopt a fictional scenario and place it in the real-life context of waste in Buenos Aires.

Jerry Diamond, an engineer, working for the UK branch of the multinational company *WasteSolutions* has been commissioned to travel to Buenos Aires to set up a new sorting and recycling facility for plastic waste in Buenos Aires. The factory would be situated next to the only remaining functioning landfill site in BsAs, *CEAMSE*. CEAMSE has a commercial interest in the project because they currently receive money for every Kg of waste they place in the landfill. It is not in their interest to recycle unless they can profit from the new factory in some way. You have been told by local university professors that you need to be careful of their connections to organised crime. The local government is keen to find solutions for the reduction of waste and have been working with *Greenpeace* to create the 'zero garbage law.' They also wish to solve what they call the *'problem of the cartoneros,'* the informal rubbish pickers who scavenge for recyclable goods to sell. When Jerry arrives in BsAs, he finds that sorting units already exist on the site. These are known as *social factories* and are run by local groups of cartoneros. The government has commissioned five other sorting units, which are run by cooperatives of cartoneros. Groups running the social factories and the sorting units keep any profit they make from sorting and selling but do not get a wage. Then there are the 10-20,000 street and cooperative cartoneros who collect, sort and, in some cases, recycle waste to make a living. This is more complicated than Jerry had expected. He decides to send for a team from WasteSolutions to come down to assess the situation with him. He sets up meetings with representatives of all known stakeholders.

Your group will take the role of one of these sets of stakeholders and in order to prepare for your meeting, you must decide, using the Franklin model, what the arrival of the new factory will do to benefit you or cost you, socially, environmentally and economically.

- Street cartoneros

- CEAMSE

- Waste Solutions

CHAPTER 8

Student Involvement

8.1 STUDENT INVOLVEMENT IN WFL BUENOS AIRES

At the time of writing, WFL BsAs and Lesotho have involved over 80 project students internationally. To date, the following students have been involved: 1 geography student studying the role of the government with cooperatives (UCL, UK); 1 recent business graduate working on cost benefit analysis (Germany); 24 first year engineering project students designing products (Queens, Canada); 7 undergraduate engineering thesis students on materials development (Queens Canada); 2 PhD students on materials development (U. of Naples, Italy, Queens, Canada); 6 engineering design project students (Queens, Canada); 5 social science, developlent studies and engineering project students working on product design and feasibility (Queens, Canada); 1 class of design student on product design in the US (RISD); and one in Buenos Aires (UBA, FADU); 1 film/development studies student on WFL BsAs documentary (Queens, Canada); 1 Latin America studies student helping with translation (U of Alberta, Canada); 3 mechanical and one environmental engineering thesis students; and one team of first year students (UWA, Perth).

Maintaining the focus and values of the work across such a broad swath of disciplines and countries has presented yet more challenges to the group. We use online networking technologies to keep us all in touch, including key contacts in BsAs, and the growing international support network of students has become a critical component of the success of WFL Buenos Aires. International students will visit BsAs and join the local students in working with the cooperatives, but only when we are certain that there will be mutual benefit in doing so. In the meantime, student support has enabled the project to develop more rapidly than would have been otherwise possible.

As a result of the students support so far, we have been able to explore the properties of the materials so that we can determine which applications they are suitable for. We have also explored different waste material combinations, different amounts of plastic versus fibre, different fibres, different plastics from a variety of sources, as well as varying processing techniques. For all of these combinations, we have studied the strength, toughness, stiffness, creep resistance, moisture resistance and flame resistance of the materials (Thamae and Baillie, 2007, 2008; Thamae et al., 2009a,b).

Many students have developed product prototypes based on the design specification brief of WFL, adopting the idea of cooperatives of cartoneros creating products to sell in Buenos Aires from locally found waste plastic and either cardboard, paper or textiles. Engineering students from Queens (first, final year and graduate students), as well as design students from UBA and RISD have created a series of potential products. Some of the Queens students' designs and all of the UBA and RISD student designs have been manufactured. These are currently being assessed for their viability as products for different cooperative groups, by the cartoneros themselves relating to materials sourcing and practicality or manufacture and for marketing issues as discussed in Chapter 7. Some of the designs are given in Figure 8.1 below.

Figure 8.1: Designs produced by students for Waste for Life Buenos Aires.

8.2   REFLECTIONS ON STUDENT INVOLVEMENT

Service learning in engineering is becoming increasingly popular, especially in North America, e.g., Engineering Projects in Community Service (Coyle et al., 2005) and organisations such as Engineers without Borders are rapidly appearing in the US, Canada, Australia and the UK (Kabo and Baillie, 2009a). However, despite the worthy intentions of a few academics, it is clear that in many cases, students are not sufficiently prepared for community service, and that it is, in fact, students who are benefiting more than the communities they intend to serve (Vandersteen et al., 2009).

This is not a new phenomenon for service learning in general. In 2000, Marullo and Edwards (2000, p. 899-906) discussed service learning as one way for universities to form collaborative partnerships with the community to address social, political, economical, and moral ills. They clearly point out the difference between work that aims to promote social justice and work which is done in a charity framework. While social justice aims to change an unjust structure, charity, whilst necessary and important, provides only a temporary solution that often ends up reproducing the status quo rather than challenging it. Marullo and Edwards stress that it is essential to ask who is empowered by work undertaken by students and whether the work does anything to address the root causes of the problem in question. Bringle and Hatcher further discuss this in relation to the university 'expert model.'

> Morton (1995) observes that campus-community partnerships are too often rooted in charity rather than justice. Charity occurs when resources and surplus are given from one community to another community, whereas justice is demonstrated when resources are considered as mutual resources and shared among members of the same community....Benson, Harkavy and Puckett (2000) note that the expert model, which is frequently used by faculty members, is one in which the relationships are elitist, hierarchical, and unidimensional rather than collegial, participatory, cooperative, and democratic (Bringle and Hatcher, 2002, p. 14).

In engineering fields, these questions are largely unexplored, if not completely ignored, partly because of the disciplinary silo in which we work and partly because of our special relationship with entrenched power structures. Students and even engineering professors are not encouraged to critique and, when they do, they come up with socially based questions which they do not have the skills to problematize, so questions of who benefits and who is harmed and what reinforces those imbalances can go unchallenged.

The peculiar decoupling of the learning potential of service learning projects from the project itself as well as the potential harm this may cause within communities, is our main concern. In social science and humanities disciplines, it is usual for students to learn to deconstruct arguments by asking who makes the statements, why are they making them, why are they being made here and now, who are the beneficiaries, and at whose and at what expense? Historians and development studies students, in particular, learn how to think critically and historically, in order to learn from the past. For some reason, this is absent from engineering studies. In the rare case that history is mentioned, it is restricted to the lessons learnt about communication or technical errors when, for instance, a bridge has collapsed. The broader implications of the ways in which science and technology influence and are influenced by the values of a society (Bijker, 1995, 1998) are not usually explored except in the very few universities which offer courses from a science and technology studies unit, e.g., Rensellear Polytechnic Institute (Pritchard and Baillie, 2006). Thus, engineering students, who embark on community service projects, rarely have access to systematic preparatory or post education courses that focus on the cultural and social contexts of the work, resulting in potentially hazardous tensions with the communities they are intended to serve.

How then do we help students think through their work and, especially, their service learning work with heightened sensitivity to the many pushes and pulls in a development project, which by its very nature implies disparities of power and privilege? It is possible to use audio, image, and video enhanced presentations to point out the complexities of any given situation, explain the variety of socio-political framing, and the viewpoints and experiences of different constituencies. But how is it possible to prepare engineering students to step outside of their own 'thought collectives' so that they can critique and question their own practice in the light of this knowledge?

Richard Day and Caroline Baillie have developed a course entitled *'Engineering and Social Justice'* (Kabo and Baillie, 2009a,b), which is intended to help students develop 'critical' thinking in relation to technology engineering practices. It locates the achievements and failures of technological rationality as it has been deployed within European modernity, both historically and in its current phase of neoliberal globalisation. It also explores alternative ways of relating to technology that are non-exploitative, non-oppressive, and ecologically sustainable. The course is cross-disciplinary both in content and student composition (engineering and social science). Social justice is the guiding ideal of the course and how it is relevant in an engineering context are explored in multiple ways. The classes are carried out as seminars with discussions usually centred on assigned weekly readings. These readings present the students with a range of perspectives and cover key concepts, the dominant engineering paradigm, critical perspectives and alternative paradigms. The students are required to carry out community based group projects in which they critically examine an element of engineering practice and in which they can begin to explore their own biases and habits of thought. An important element of the course is the presence of both instructors at all times, which allows students to hear disagreements and differences between two professionals and to show them that this is OK. The course has been researched with the question 'can we help students pass through the threshold and see engineering through a lens of social justice?' Findings from this study have demonstrated that it is in fact possible to help students critique their own practice in ways which lead to a more knowledgeable and 'response- able' approach (able to respond) (Kabo and Baillie, 2009a,b). All Queens University students involved in WFL projects have been exposed to the course or key elements of it in one-to-one or small group sessions, and it is recommended that short intensive versions of this kind of course be made available to students before embarking on any service learning project.

Before involving students in any form of service learning, we need to consider some key questions:

1) At which point should students get involved? Can they be involved from the begin-ning/exploratory phase or do they need to wait until after this phase is complete?

   We took on two students after about three months in BsAs. Our experience working with them was very positive, but it took much one-on-one advising, on-the-street preparation, and sensitivity training for this to be possible. If such time is available, and especially for higher degree students, this can be an important element of team building as it brings in diverse perspectives, especially if students of different backgrounds (disciplines, nationality, culture) are involved. Otherwise, we would suggest that it is better to bring in students at a later phase.

2) In order to involve themselves in this kind of work, should students be enrolled in a particular class or can this be done as an individual project?

   We have had involvement with students in many different contexts and would only say that it is important that the level of involvement matches the level of students understanding of the context and support from the university and their professor. For example, one-to-one support for projects is great so long as the professor is giving adequate appropriate support. Large classes are also fine but only if the students involvement remains at a distance for partners on the ground, e.g., the first year students offering product design solutions.

3) Should students stay at home, work in the field, or?

One of the key questions to ask is whether students actually need to do field work in order to support these programs. In our case, apart from the two students mentioned above, all students remained in situ in their own universities during the set-up stage. The negative consequence of this is the reduction in the excitement and potential learning experience of students engaging with the community partners and developing a better understanding of their cultures. However, this does not mean that no valuable service learning can take place.

We are developing an emerging methodology for preparing engineering students to work within a service learning paradigm. We critique engineering as a culture whose 'common sense' does not fit well with progressive, participatory practices in developing countries and work to create an approach necessary for the preparation of student involvement whilst planning and implementing an effective multiple partner network and program.

The elements include:

Project Development, Identifying Networks and Exploring Dynamics
1) Students can get involved at all stages of the project development in order to learn how to manage and understand the complexity of decisions that need to be made. Students can support the preparation of the project in situ in their own universities and become involved in the project design cycle (based on a modified Catalano model) from the start.

Core Values
2) Education of students for involvement of any stage of the project should involve the development of a critical lens, promoted by interactive, interdisciplinary seminar style discussions on key social and political frameworks with which to critique engineering practice in a development context and consider alternatives to 'normalized' engineering paradigms. This could also be offered in one-to-one support by a project supervisor.

Documenting and Critiquing
3) Students and team members should be kept informed and in touch using a variety of networking technologies, also by carefully documenting the project development using video where possible.

Timing and location of student involvement
4) Involvement of students in the preparatory stages of problem definition and project development is best at a distance and not in direct contact with community members at this delicate

initiation state. At this stage, community team members begin to identify their own needs and to realize benefits, and cultural tensions within communities are being explored.

5) Involvement of a network of international students is encouraged, both local to the project as well as in other continents, whose viewpoints will compliment one another and form a global team.

6) Students can get involved directly with community project partners and begin to do field work immediately if there is enough one-to-one supervision on the ground and language barriers have been addressed. Otherwise, they can become fully involved after the above preparatory stages have been completed.

CHAPTER 9

Summary Thoughts

This book summarises the story of Waste for Life and its emergence in two very different contexts, in order to provide an in-depth case based guide to community development for the neophyte engineer. We have chosen to share the early stages of intervention with each community and show how we began the journey. Too often we hear about the results of development projects, the successes or failures, but we rarely learn how the relationship began. This is the story of that beginning and of how we got involved in these two very different contexts. There are those who would suggest that we should stay away, that we should not continue our colonial past, that Northern engineers should not support Southern communities. Whilst we have sympathy with the very important reasons for this argument, we also believe that engineering is a potentially powerful way of supporting socially just and sustainable development, which is mutually negotiated with local partners on the ground. Sharing, or, as the cartoneros would say, socializing our knowledge with marginalized groups, seems to be a critical and necessary future action for all engineers, North or South. As such, we need to understand what we are doing and why. We must not take on the trustee mentality of the past. We must not think ourselves superior by our actions. We must listen, we must observe and we must learn to see what our technical knowledge can bring to the community, in ways that will support positive, life-giving structures as well as independence from unsympathetic political systems.

We hope that the issues raised in this book will provide the guidance that we intend – not as a how-to manual – but as a framework for planning projects, for engaging with communities, and mostly, in order for others to learn from our mistakes, to learn from our ignorance. We wish to provide a safety net for student engineers whose dream it is to use their skills to build a better future.

Bibliography

Baillie, C., (2002). Negotiating Scientific Knowledge in Entangled histories and negotiated universals: Centres and peripheries in a changing world, Campus Verlag, Berlin 2002, pp. 32–57.

Baillie, C., (2003). Why Green Composites? in Baillie C. Green Composites: Polymer Composites and the Environment. Cambridge, Woodhead Publishing Limited.

Baillie, C. and Catalano, G., (2009). Engineering and society: working towards social justice, Morgan & Claypool Publishers. DOI: 10.2200/S00137ED1V01Y200905ETS009 7

Baillie, C., (2006). Engineers within a local and global society, Morgan & Claypool Publishers, San Rafael. DOI: 10.2200/S00059ED1V01Y200609ETS002 1, 5, 13

Baillie, C. and Catalano, G., (2009). Engineering and Society: working towards social justice: Parts 1, 2 and 3, Morgan & Claypool Publishers. DOI: 10.2200/S00059ED1V01Y200609ETS002

Baillie, C. and Rose, (2004). 4

Benson, Harkavy and Puckett, (2000) 105

Berger, G. and Blugerman, L., (2006). Argentina: Recover Them from Oblivion. Recover the Community's Ability to Produce. Cristina Lescano and El Ceibo. Revista: Harvard Review of Latin America, Autumn Issue. 58, 61, 65

Bijker, (1995). 105

Bijker, (1998). 105

Brawley, (0000). 3

Bringle, R.G. and Hatcher, J.A., (2002). **Journal of Social Issues**, Sep 2002, Vol. 58, Issue 3, pp. 503–516, p. 14. 105

Bullentini, A., (2008). "Cartoneros y Monotributistas."
http://www.pagina12.com.ar/diario/sociedad/3--114312-2008-11-01.html 94

Burton, (2005). Despite the country's large urban population, Argentina's cities are given few real powers. City Mayors [Online]. Available from
http://www.citymayors.com/government/argentina_government.html
[Accessed January 2008].

Capra, (2002). 2, 3

Catalano, G., (2006). Engineering ethics, peace justice and the earth Synthesis lectures on Engineers, Technology and Society, (Ed. Baillie, C.) Morgan & Claypool. DOI: 10.2200/S00039ED1V01Y200606ETS001 5, 6, 7

Catalano, G.D., (2007). Engineering, poverty, and the earth (C. Baillie, Ed.), Synthesis lectures on Engineers, Technology and Society #4. San Rafael, California: Morgan & Claypool Publishers. DOI: 10.2200/S00088ED1V01Y200704ETS004 5

CEAMSE, (2007). Estudio de calidad de los residuos solidos urbanos. CEAMSE Enero. 48

Chapeyama, O., (2004). Development of a landfill in Maseru, Lesotho : Report on assessment of needs for landfill development, USAID, http://www.dec.org/pdf_docs/PNADB395.pdf. 11

Chen, M., (2007). Rethinking the Informal Economy: Linkages with the Formal Economy and the Formal Regulatory Environment. United Nations Department of Economic and Social Affairs Working Paper No. 46. [Online] Available from
http://www.un.org/esa/desa/papers/2007/wp46_2007.pdf
[Accessed January 2008]. 55, 70, 71, 73

Chen, M., Jhabvala, N., and Lund, F., (2001). Supporting Workers in the Informal Economy: A polic Framework. Policy Paper for the International Labour Organisation Task Force on the Formal Economy. 65, 68

Chevalier, J.M. and Buckles, D.J., (2008). SAS2 A guide to collaborative enquiry and social engagement, Sage. 47

Chronopoulos, T., (2006). Neo-liberal reform and urban space; the cartoneros of Buenos Aires, 2001–2005. City, 10(2). Taylor & Francis: London. DOI: 10.1080/13604810600736651 56, 58, 61, 64, 75

Cibils, A., (2006). Till Debt Do Us Part: Lessons from Argentina's Experience with the IMF, Debt, and Financial Crises. Americas Program: USA. 64

CNN (2003). Accommodating an army of garbage pickers. CNN.com. [Online] Available from   http://www.cnn.com/2003/WORLD/americas/03/26/argentina.train.reut. [Accesses September 2007]. 60

Cobbe, J., (1982). Migrant labour and Lesotho: Problems and policies and difficulties and constraints with respect to them, The Institute of Labour Studies Occasional Paper, Maseru, Lesotho.

Coburn, J., (2005). Street Science: community knowledge and environmental health justice. Cambridge, MA: MIT Press.

Corbiere-Nicollier, T., Laban, B.G, Lundquist, L., Leterrier, Y., Manson, J.A.E., and Jolliet, O., (2002). Lifecycle assessment of biofibers replacing glass fibers as reinforcement in plastics, Resource Conservation Recycling, 33(4), pp. 267–287. DOI: 10.1016/S0921-3449(01)00089-1

Cowan, R. and Shenton, R., (1996). Doctrines of Development, Routledge. 28

Coyle, E.J., Jamieson, L.H., and Oakes, W.C., (2005). EPICS: Engineering Projects in Community Service. International Journal of Engineering Education, 21(1), pp. 139–150. 104

Crooks, H., Lorimer and company, (1993). Giants of Garbage: The Rise of the Global Waste Industry and the Politics of Pollution Control, Lorimer and company. 50, 51

Daniel, J. and Rosen, M., (2002). Exergetic environmental assessment of life cycle emissions for various automobiles and fuels, Exergy, An International Journal, 2(4), pp. 283–294. DOI: 10.1016/S1164-0235(02)00076-6

Darder, Baltodano and Torres, (2009). 4

Defourny, J. and Develtere, P., (1999). The Social Economy: A Worldwide Making of a Third Sector. In: Defourny, J., Develtere, P., and Fonteneau, B., Eds. L'economie Sociale au Nord et au Sud. De Boeck University: Belgium. 55, 56, 61

De La Torre, A., Yeyati, E., and Schmukler, S. (2002). Argentina's Financial Crisis: Floating Money, Sinking Banking. World Bank. [Online] Available from http://wbln0018.worldbank.org/LAC/lacinfoclient.nsf/1daa4610322912388525683100 5ce0eb/74fec6924c42c4ed85256c22005cc4bd/FILE/Floating%20Money%20Sinking%20Banking%20(3Jun02).pdf, [Accessed November 2007]. 64

Deleuze, G. and Guattari, F., (1987). A Thousand Plateaus, Capitalism and Schizophrenia, Uni Minnesota Press. 7

De Soto, H., (1989). The Other Path. Tauris: London. 71

Dey, I., (1993). Qualitative Data Analysis: A User-Friendly Guide for Social Scientists. Routledgte: London. 57

Diener, J. and Siehler, U., (1999). Okologischer vergleich von NMT-und GMT-Bauteilen, Angew Makromol Chem, 272, Nr. 4744. DOI: 10.1002/(SICI)1522-9505(19991201)272:1%3C1::AID-APMC1%3E3.0.CO;2-4 99

Fabiani, A., (2005). "Cooperativa El Ceibo: De cirujas a empresarios." Extramural. Fall 2005. Nexus. http://extramuros.unq.edu.ar/03/art_coop_ceibo_3.htm. 94

Factbook, (2009). The CIA World Factbook. Washington, DC: Central Intelligence Agency. https://www.cia.gov/library/publications/the-world-factbook/index.html. 11

Feldstein, (2002). Argentina's Fall: Lessons from the Latest Financial Crisis. Foreign Affairs. March/April 2002.

Ferguson, J., (1990). The Anti-politics Machine: 'Development,' Depoliticization and Beurocratic Power in Lesotho, Cambridge: University Press. 4, 12, 77

Fleck, L., (1979). Genesis and Development of a Scientific Fact. Chicago: University of Chicago Press. 4, 5

Franklin, U., (1999). The Real world of technology, House of Anansi Press.

Garcia, M.J., (2007). "The White Train." The Virginia Quarterly Review, pp. 38–61. 94

GCBA, (2008). Gobierno de la Ciudad de Buenos Aires. Available from http://www.buenosaires.gov.ar/ [Accessed January 2008]. 63, 64, 67, 68, 69

Goedkoop and Oele, (2004). Introduction to LCA, Simapro. 99

Gramsci, A., (1971). Selections from the prison notebooks of Antonio Gramsci. (Hoare.) 4

Gupta, A., (1999). Science, sustainability and social purpose: barriers to effective articulation, dialogue and utilization of formal and informal science in public policy. International Journal of Sustainable Development. Vol 2, No. 3. 77

Hamer, J.H., (1981). Limits in the Formation of Associations: The Self-Help and Cooperative Movement in Sub-Saharan Africa. African Studies Review. 24(1), pp. 113–132. DOI: 10.2307/523914

Hart, K., (1973). Informal Income Opportunities and Urban Employment in Ghana. The Journal of Modern African Studies. 1(1), pp. 61–89. DOI: 10.1017/S0022278X00008089

Hay, R.K.M., (1995). Harvest Index: a review of its use in plant breeding and crop physiology, Annals of Applied Biology, 126(1), pp. 197–216. DOI: 10.1111/j.1744-7348.1995.tb05015.x 37

Hoare, Q. and Smith, G.N., (1971). The Study of Philosophy: Introduction. In Gramsci, A., Hoare, Q., and G. Nowell Smith, Eds., Selections from the prison notebooks of Antonio Gramsci, pp. 321–322. New York: International Publishers. 4

Hull, D., (1990). An introduction to composite materials, Cambridge, University Press, Cambridge. 35, 36

iEco, (2009), Estiman que los ingresos turísticos caerón un 7% este año. Buenos Aires: 19 Oct 09. http://www.ieco.clarin.com/economia/ Estiman-turisticos-ingresos-caeran-ano_0_68700014.html. 97

Jeter, J., (2003). "Scrap by Scrap, Argentines Scratch Out a Meager Living." The Washington Post. 7 June 2003.

Johnston, S.F., Gostelow, J.P., and King, W.J., (2000). Engineering and Society. Toronto: Prentice-Hall Canada Inc. 3

Joshi, S.V., Drzal, L.T., Mohanty, A.K. and Arora, S., (2004). 'Are natural fiber composites environmentally superior to glass fiber reinforced composites?' Composites Part A: Applied Science and Manufacturing, 35(3), pp. 371–376. DOI: 10.1016/j.compositesa.2003.09.016 99

Jossa, B., (2005). Marx, Marxism and the Cooperative Movement. Cambridge Journal of Economics, 29(1), pp. 3–18. 55, 56

Kabo, J. and Baillie, C., (2009a). Seeing through the lens of social justice: a threshold for engineering in Meyer, J., Land, R., Baillie, C., Eds., Threshold concepts and transformational learning Sense Publishers. 3, 104, 106

Kabo, J. and Baillie, C., (2009b). Engineering and social justice: Negotiating the spectrum of liminality, to appear in Baillie, C., Bernhard, J., Eds., Special issue Engineering Education Research European Journal Engineering Education (Vol. 34, issue 4). 106

Krozer, J. and Vis, J.C., (1998). How to get LCA in the right direction, Journal of Cleaner Production, 6(1), pp. 53–61. DOI: 10.1016/S0959-6526(97)00051-6

Leduka, (2006). 12

Leftwich, A., (2000). States of Development, Polity Press. 1

Lesotho Bureau of Statistics, (2001). Lesotho demographic survey, Maseru, Lesotho. www.bos.gov.ls. 15

Lesotho Bureau of Statistics, (2002). Lesotho demographic survey, Maseru, Lesotho. www.bos.gov.ls. 16, 29, 30, 31

Lesotho Bureau of Statistics, (2002). Lesotho environmental statistics, Maseru, Lesotho. www.bos.gov.ls.

Lesotho Bureau of statistics, (2002). Lesotho agricultural situation report: 2000/01–2001/02, Maseru, Lesotho. www.bos.gov.ls.

Levy, D., (2007). Review: The Other Path: The Invisible Revolution in the Third World. Review for Federal Reserve Bank of Minneapolis. [Online]. Available from http://www.mpls.frb.org/pubs/region/89--12/review.cfm. [Accessed February 2008].

Li, T.Q. and Wolcott, M.P., (2004). Rheology of HDPE- wood composites: stead y state shear and extensional flow, Composites Part A Applied Science and Manufacturing, 35(3), pp. 303–311. DOI: 10.1016/j.compositesa.2003.09.009 99

Lucena, J., Schneider, L.J., (2010). Engineering and Sustainable Community Development, Morgan & Claypool. 7

Makoa, F.K., (1999). Lesotho's rural development policy: Objectives and problems, in Review of African Studies, 3(1), p 38–60. 12

Marake, M., Mokuku, C., Majoro, M., and Mokitimi, N., (1998). Global change and subsistence rangelands in Southern Africa: Resource variability, access and use in relation to rural liveli-hoods and welfare. The National University of Lesotho, Roma, http://www.bangor.ac.uk/rangeland/reports/le-task0.htm. 15

Marcouillier, D., Ruiz de Callista, V. and Woodruff, C., (1997). 'Formal Measures of the Informal Sector Wage Gap in Mexico, El Salvador and Peru.' Economic Development and Cultural Change. 55

Marullo, S., and Edwards, B., (2000). From Charity to Justice: The Potential of University-Community Collaboration for Social Change. American Behavioral Scientist, 43, pp. 895–912. DOI: 10.1177/00027640021955540 104

Mayoux, L., (2006). Participatory Methods. Occasional Paper. [Online] Available from. http://www.enterprise-impact.org.uk/pdf/ParticMethods.pdf. [Accessed December 2007].

Mendieta, E., (2006). The Future of the Past: Latin American Cities Yesterday and Today. City, 10(2). DOI: 10.1080/13604810600869064 56

Medina, M., (2000). Scavenger Cooperatives in Asia and Latin America. Resources. Conservation and Recycling, 31. DOI: 10.1016/S0921-3449(00)00071-9 58

Medina, M., (1997). Informal recycling and collection of solid wastes in develop-ing countries: Issues and Opportunities, July 1997, UNU/IAS Working Paper 24. DOI: 10.1016/j.habitatint.2008.05.006 50

Medina, M., (2005). Waste Picker Cooperatives in developing countries. Presented at the Wiego/Cornell/SEWA Conference on Membership-based organisations of the poor, Ahmed, India, 2005. 48, 49

Morton, (1995). 105

Murphy, R., (2003). Life Cycle Assessment, In Baillie, C, Green Composites: Polymer Composites and the Environment, Cambridge: Woodhead Publishing Limited. 99

Mvuma, K.G.G., (2002). Urban poverty reduction through municipal solid waste management (MSWM): A case study of Maseru and Maputsoe in Lesotho, A PhD thesis submitted in the Department of Civil Engineering, University of Durban-Westville, Durban. 11, 30, 32

Nas, P. and Jaffe, R., (2004). Informal Waste Management: Shifting the Focus From Problem to Potential. Environment, Development and Sustainability, 6, pp. 337–353.

Neamtan, N., (2002). The Social and Solidarity Economy: Towards an 'Alternative' Globalisation. Conference Paper for 'Citizenship and Globalisation; Exploring Participation and Democracy in a Global Context.' 61

Pirez, A., (2002). Buenos Aires: fragmentation and privatisation of the metropolitan area. Environment and Urbanisation, 14:145. Sage. DOI: 10.1177/095624780201400112 61

Portes, A. and Schauffler, A., (1993). Competing Perspectives on the Latin American Informal Sector. Population and Development Review, 19(1), pp. 33–60. DOI: 10.2307/2938384 66

Prihar, S.S., and Prihar, B.A., (1991). Stewart sorghum, Harvest Index in relation to plant size, environment, and cultivar, Agronomy Journal, 83, pp. 603–608. 36

Pritchard, J. and Baillie, C., (2006). How can engineering education contribute to a sustainable future? European Journal of Engineering Education, Vol. 31, No. 5, pp. 555 – 565, October 2006. DOI: 10.1080/03043790600797350 105

Rakowski, C.A., (1994). Contrapunto: The Informal Sector Debate in Latin America. SUNY Press: New York.

Riley, D., (2008). Engineering and Social Justice. San Rafael, California, Morgan & Claypool Publishers. DOI: 10.2200/S00117ED1V01Y200805ETS007 5

Rothschild, J. and Russell, R., (1986). Alternatives to Bureaucracy: Democratic Participation in the Economy. Annual Review of Sociology, 12, pp. 307–328. DOI: 10.1146/annurev.so.12.080186.001515

Schamber, P.J. and Suarez, F.M., (2007). Recicloscopio, Miradas sobre recuperadores urbanos de residuous de America Latina, Prometeo Libros. 48

Schmidt, W. and Beyer, H., (1998). Life cycle study on a natural-fiber-reinforced component, Society of Automotive Engineers, SAE technical paper series, 982195.

SETAC, (1993). Guidelines for life-cycle assessment: a 'code of practice,' from the SETAC Workshop held at Sesimbra, Portugal, 31 March-3 Pensacola, FL: Society of Environmental Toxicology and Chemistry: SETAC Foundation for Environmental Education. 99

Sclove, (1995). 77

Sen, A., (1999). Development as Freedom, Anchor Books. 2, 100

Sivertsen., L.K., Haagensen, O.J., and Albright, D., (2003). A Review of the Life Cycle Environmental Performance of Automotive Magnesium Society of Automotive Engineers, SAE Technical Paper Series, no. 2003–01-0641. 99

Smith, G.N., Ed., New York: International Publishers.

Smith, M.P., (1988). Power, Community and the City. Transaction Publishers: USA. 55, 56

Suarez, J., (2003). In Reuters: Accommodating an army of garbage pickers. [Online] Available from http://www.cnn.com/2003/WORLD/americas/03/26/argentina.train. reut/. [Accessed January 2008].

Thamae, T. and Baillie, C., (2008). A life cycle assessment of wood based composites: A case study, In Oksman K., and Sain, M., Wood-polymer composites, Woodhead Publishing Limited, Cambridge. 99, 103

Thamae, T. and Baillie, C., (2007). Influence of fibre extraction method, alkali and silane treatment on the interface of Agave americana waste HDPE composites as possible roof ceilings in Lesotho Composite Interfaces, Volume 14, Numbers 7–9, pp. 821–836 (16). DOI: 10.1163/156855407782106483 103

Thamae, T., Marien, R., Wu, C., Chong, L., and Baillie, C., (2008). Developing and Characterizing New Materials Based on Waste Plastic and Agro-fibre, Journal of Materials Science, Vol. 43, No. 12, pp. 4057–4068. DOI: 10.1007/s10853-008-2495-3

Thamae, T., Aghedo, S., Baillie, C., and Matovic, D., (2009a). Tensile Properties of Hemp and Agave Fibres in: Bunsell and Schwartz, Eds., Handbook of Tensile Properties of Textile and Technical Fibres, Woodhead Publishing, UK. 103

Thamae, T., Vaja, S., Shangguan, Y., Finoro, C., Stefano, N., Baillie, C., (2009b). Mechanical and moisture absorption of corn and wheat flour composites for developing countries, in Green Composites: Properties, Design and Life Cycle, Willems, F., and Moens, P., Eds. 103

Turkon, D., (2003). Modernity, Tradition and the demystification of cattle in Lesotho, African Studies 62, 2, Dec., pp. 147–168. DOI: 10.1080/0002018032000148731

UNEP, (1996). (United Nations Environment Programme) Life Cycle assessment: 'what is it and how to do it,' Paris, UNEP. 99

UST, (2007). Informal publication. 61

Vandersteen, J., Baillie, C., and Hall, K.R., (2009). International humanitarian engineering placements: who benefits and who pays? IEEE Technology and Society 2009. Special issue on Volunteerism and Humanitarian Engineering, Part 1, Vol. 28, No. 4, Winter 2009. 104

Vesilind, (2006). 3

Wenger, (1998). 5

Wotzel, K., Wirth, R., and Flake, R., (1999). 'Life cycle studies on hemp fiber reinforced components and ABS for automotive parts,' Angew Makromol Chem, 272(4673), pp. 121–127. DOI: 10.1002/(SICI)1522-9505(19991201)272:1%3C121::AID-APMC121%3E3.0.CO;2-T 99

Authors' Biographies

CAROLINE BAILLIE

Caroline is a Professor of Engineering Education and Materials Engineering at the University of Western Australia as well as being co-director of Waste for Life. Caroline has published more than 180 papers and books and is cofounder of the Engineering, social justice, and peace network (esjp.org).

ERIC FEINBLATT

Eric is co-director of Waste for Life as well as a photographer, film maker, and educational consultant, specializing in online cross-cultural collaborative learning for the State University of New York.

THIMOTHY THAMAE

Thimothy is a lecturer in the University of Lesotho and founder of Waste for Life Lesotho.

EMILY BERRINGTON

Emily is a student, actor, and singer. Her thesis work contributed to Waste for Life Buenos Aires.

Printed in the United States
by Baker & Taylor Publisher Services